THE YOWIE FILE
ENCOUNTERS WITH AUSTRALIAN APE-MEN

THE YOWIE FILE
ENCOUNTERS WITH AUSTRALIAN APE-MEN

TONY HEALY & PAUL CROPPER

ILLUSTRATIONS BY "BUCK" BUCKINGHAM
COVER ART BY BARRY OLIVE

Fortean Publishing
Sydney, Australia

First published in 2023 by Fortean Publishing

978-0-6457392-0-6 (Paperback)
978-0-6457392-1-3 (Hardback)
978-0-6457392-2-0 (eBook)

Cataloguing-in-publication data is available from the National Library of Australia

Cover design and typesetting by Odyssey Publishing, www.odysseypublishing.com

This book is dedicated to our friend Dean Harrison,
founder of Australian Yowie Research (AYR).

Dean has shared hundreds of eyewitness reports with us over the years. Without his unstinting generosity, *The Yowie File* would have been a very much slimmer volume.

We're confident that this collection of cases contains enough information to convince most readers that the yowie phenomenon is more than just an amalgamation of myth, mass hysteria and hoaxes.

We encourage those still harbouring doubts to visit yowiehunters.com, where they can listen to the testimony of hundreds of eyewitnesses recorded by Dean and his team since the mid-1990s.

Throughout the years many other people have assisted us with information, encouragement and advice. We would particularly like to thank the following:

Joe Beelart, Bob Bartholomew, Sarah Bignell, Buck Buckingham, Paul Braxton, Pam Bryant, Steve Carter, Bill Chalker, Paul Clacher, Andre Clayden, Ray Crowe, Jon Downes, Roger Frankenburg, Neil Frost, Lyall Gillespie, Tony Harris, Peter Hassall, Geoff Healy, David Hearder, Graham Joyner, Wayne Knight, Rebecca Lang, Kewaunee Lapseritis, Bernie Mace, Chris Murphy, Gary and Carmel Opit, Dan Perez, Thom Powell, Tohby Riddle, Patricia Riggs, Steve Rushton, Malcolm Smith, Percy and Steven Trezise, Cecil Thompson, Tim the Yowie Man, Graham Walsh, Meryl Tobin, Mike Williams and David Window.

The book's title, *The Yowie File*, is a "tip of the hat" to our late friend, the veteran sasquatch researcher John Green. (One of John's many books about the mysterious hairy giants of North America is titled *The Sasquatch File*).

CONTENTS

INTRODUCTION

In *The Yowie* (Anomalist Books, 2006), we documented everything we then knew about Australia's mysterious Hairy Man – a yeti-like or sasquatch-like creature that figures prominently in Indigenous Australian lore and has been encountered by many European Australians since the early 1800s.

We chronicled the apparently age-old mystery from the pre-colonial era to 2006, presenting every eyewitness report and every fragment of relevant Indigenous lore of which we were then aware.

We discussed physical traces associated with yowie activity – footprints, droppings, "yowie beds", and tree damage, and described the overwhelming fear the creatures trigger in other animals, including humans. We also showed, by way of statistical analysis, that there appear to be seasonal patterns to yowie activity (or, at least, to yowie reports).

The Yowie File isn't intended to be a full-fledged sequel to *The Yowie*. It is primarily a casebook – a collection of eyewitness reports to complement the 300 we documented in the earlier book.

While each item tells its own story, we have added a few brief comments to draw attention to geographical patterns and to highlight some odd, but possibly significant, details that crop up time and again in eyewitness testimony in widely separated parts of the continent.

Golden oldies

We have always particularly valued reports from the colonial (1788-1900) and early modern (1901–1974) eras, when people who encountered the strange creatures were far less likely than contemporary witnesses to have their testimony influenced by something they read.

In the seventeen years since *The Yowie* was published, we have unearthed many more of those older reports – reports we think of as nuggets of "cryptozoological gold."

To prevent those long-lost gems from becoming scattered again, we have blended them with all of the pre-1975 reports we knew of prior to 2006 and packaged them, with a minimum of editorial comment, in what we hope will be a handy sourcebook.

We hope that by detailing the long history of Hairy Man sightings from the early colonial era onwards, we will finally put paid to the notion that the phenomenon is simply the result of fantasy-prone Australians ape-ing (so to speak) their bigfoot-hunting American cousins. That way of "explaining away" the yowie phenomenon never really passed muster because, before the very late 1960s, Australian print media featured virtually nothing about the sasquatch/bigfoot, and television documentaries about the American mystery didn't appear on our screens until the mid-to-late '70s.

Thoroughly modern yowies

While we treasure the very early yowie stories, we also value contemporary eyewitness reports that illustrate the ongoing nature of the phenomenon.

Dean Harrison of AYR is the most high-profile and active researcher of the modern era. In 1997, motivated by his own hair-raising encounter (*The Yowie*, pp. 85-87), he set up a website, yowiehunters.com, where others could report similar experiences.

It was a game-changer, immediately taking Hairy Man research several huge, yowie-sized steps forward. Dean was amazed at how many responses he received – some from people who'd sighted mysterious ape-men decades earlier and others from people who excitedly described mind-boggling encounters they'd experienced only hours before.

Eventually, the swelling number of reports – sometimes several a week – threatened to swamp Dean and his colleagues at AYR. So we began helping out by photographing locations and conducting interviews with remotely-located witnesses.

Many of the new reports, like the freshly discovered colonial and early modern era stories, are absolute "gold", and as Dean, ever generous, agreed, we decided to include a judicious selection of them in this volume, to bring the Hairy Man story right up to date.

We hope that the additional modern-era reports (none of which appeared in our previous book) will persuade most readers that mysterious ape-men continue to be reported from one side of the continent to the other by people who, for the most part, appear to be sensible and honest. Readers may also find it significant, as we do, that many modern reports emanate from the very same areas that were notorious for "Australian ape" activity a century or more ago, and from locations Indigenous people know to be abodes of the Hairy Men.

Indigenous lore

In *The Yowie* we devoted a fourteen-page chapter to Indigenous beliefs, showing that many, if not most, Aboriginal people, from Cape York Peninsula right down the east coast to Victoria, across to South Australia, Western Australia and up to the Northern Territory, strongly believe in the existence of the creatures. Their many cultures have different terms for them – *doolagarl, thoolagarl, nooncoonah, jimbra, tjangara, puttikan* and *jurrawarra,* to name but a few. Nowadays, when discussing them with people outside their own language group, Indigenous people generally use the term "Hairy Man."

Indigenous yowie traditions vary from region to region. In many parts of the continent, it is believed that in addition to the hairy giants there exist much smaller hair-covered, man-like creatures. The little creatures are known by many names, including *junjudee, dinderi, net-net, njmbin* and *waaki.* In northern New South Wales, they are often referred to as "brown jacks" and in parts of Western Australia as "woodarchies".

As readers will discover, many non-indigenous people have also encountered the smaller variety.

Australian Aboriginal culture is so complex and varied that a comprehensive study of Indigenous Hairy Man beliefs would require a lifetime's work. (We've long thought, in fact, that there's a PhD waiting for whoever completes such a study – hopefully an Indigenous researcher.)

That being said, our files now contain considerably more Indigenous lore than they did back in 2006, and we'd like to document it as best we can.

Because of its particular format, however, we have chosen not to include a dedicated Indigenous chapter in this book. Instead, we have slotted each item of relevant lore into our catalogue of cases according to the date, or approximate date, it was documented. Most of those items fall into the colonial era, but the latter part of the book also contains considerable Aboriginal testimony, in the form of contemporary sighting reports. In documenting those incidents, we have highlighted the Indigenous identity of the eyewitnesses.

New discoveries

Recent discoveries of ancient human species such as *Homo naledi, Homo luzonensis,* Denisovans and *Homo floresiensis* (the "Hobbit") continue to challenge our understanding of human evolution.

Although some cryptozoologists think yowies might be lingering survivors of one or other of those ancient species, no known hominin seems to have closely matched the hairy giants of Australia in appearance.

Nor is there incontrovertible evidence of any hominin other than *Homo sapiens* having made the difficult sea-crossing to Australia.

Homo erectus and *Homo floresiensis* made it as far as Flores, and Denisovans may well have reached New Guinea. None of them, however, were anywhere near the size of the gigantic Aussie ape-man, and their technology (fire usage and tool-making) was far superior to anything attributed to the yowie.

Yowies – fact or fable?

We find the huge body of eyewitness testimony quite compelling, mainly because, contrary to what some sceptics suggest, at least half of all yowie encounters occur in broad daylight, and approximately one-third involve more than a single witness.

So, despite a few obviously bogus "tall stories" in the colonial-era, despite a few irritating modern-era hoaxes, and despite the likelihood that a few informants were mistaken and others deluded, we see no reason to doubt the veracity or sanity of the great majority of eyewitnesses.

But – and it's a very big, yowie-sized "but" – despite more than 45 years on the trail, we have yet to see incontrovertible evidence that the damnably elusive creatures really do exist.

A poor track record

One thing that baffles and frustrates yowie researchers, and gives heart to sceptics, is the rarity of track finds. If the Hairy Men are so massive, why are so few of their footprints found? Creatures of that size should leave many deep imprints, yet track finds feature in less than ten percent of sighting reports.

As discussed at length in 2006 (*The Yowie*, pp. 137-139, 154, 188-89), tracks that *are* found show, with a few notable exceptions, little consistency, not only in the number of toes, but also in general shape. While the five-toed variety is most commonly reported, four-toed and three-toed tracks are found almost as frequently – sometimes immediately after sightings by apparently very credible people.

Although some of the apparently yowie-related tree damage and stick formations we've examined have been very impressive, we've never seen any physical remains – not so much as a single finger bone – to prove the creatures' existence. Hard evidence has proven *very* hard to find.

Photographic evidence has also been, to put it mildly, disappointing. With the exception of some remarkable footage taken by Dean Harrison and his team in April 2021 (see Appendix), we've never seen a picture or video, purportedly of a yowie, that was of any real value.

Another good reason for doubting the physical reality of the yowie is that similar, equally elusive, creatures have been reported for centuries in many other parts of the world – the North American sasquatch, the Russian almasti, the Himalayan yeti, the Chinese yeren and the Malaysian orang mawas, to name but a few. Despite occasional unprovable claims to the contrary, none of those creatures has ever been captured or shot dead.

As late as 2006, when *The Yowie* went to press, we still thought there was a faint chance that, despite all indications to the contrary, the Hairy Men of Australia might be real, flesh and blood, albeit uncannily elusive, creatures. We certainly *hoped* so.

But as the years rolled by, the creatures continued to defy the laws of probability – evading bullets, speeding vehicles and camera traps – and what remained of that forlorn hope gradually faded.

So that left, as we saw it, only two possibilities.

One: that the yowie phenomenon is just an elaborate psychological/sociological construct – an amalgamation of myth, mass hysteria, hoaxes, misidentification of common wildlife and, just possibly, a deep-seated race-memory of our distant ancestors' conflicts with primitive hominins or hominids in Africa and Eurasia.

Because of the transparent honesty of so many informants, the many compelling multiple-witness reports, the undeniable tree damage and other physical traces, many people, particularly yowie researchers of the "flesh and blood" school, will find the psychological/sociological hypothesis totally unacceptable.

We don't much like it either, but it has to be considered.

Two: that the creatures are psychic phenomena capable of taking physical form. Interdimensional entities, perhaps. Something uncatchable and unkillable, way beyond human understanding.

That notion, too, will be unpalatable to many readers, but outrageous as it sounds, it does align well with much of the Aboriginal, Native American and Sherpa lore that has been shared with us over the years.

Throughout the 1970s, '80s, and early '90s we received a few stories that contained hints of the paranormal. Rather non-plussed, we cautiously accepted some, while consigning others to the "too weird" basket.

But now, having interviewed many more witnesses and having perused hundreds of additional cases, we have come to accept that a significant proportion of reports contains details that smack somewhat (and often considerably more than somewhat) of the supernatural. Although most remain convinced they've encountered real, flesh and blood creatures, about ten to fifteen percent of witnesses make a point of mentioning very weird details, and some firmly believe their entire experience was decidedly uncanny.

Specific "high strangeness" details crop up time and time again – strong, unsettling feelings of being watched, cessation of all bird and insect noise, levels of fear out of all proportion to the situation, post-sighting nightmares, sightings of "black panthers" and other odd creatures, floating lights and electrical malfunctions.

The same odd details have been noted many times by North American researchers including Thom Powell, Kewaunee Lapseritis, Joshua Cutchin and Timothy Renner. As our late friend Ray Crowe of Oregon, once remarked: "The more you learn about the bigfoot, the stranger it gets."

In 2005, while wrestling with this mind-bending problem, fortean researcher Jerome Clarke theorised, in the *Journal of Science Exploration*, that "... most high-strangeness reports of encounters with ... bizarre entities [might be] vividly felt but subjective anomalies of imagination and consciousness, not currently understood but presumably ... resolvable with open-minded research and expanded knowledge."

We have included all the "high strangeness" reports in this collection alongside the more numerous "normal" reports so that readers can decide for themselves if the paranormal theory has any validity. No doubt many, seeing it as just a clumsy way of "explaining" one mystery by invoking another, will think not.

We're not all that crazy about the paranormal theory, such as it is, either. But throughout our decades of research, we've always believed in going, for better or worse, where the evidence leads.

And now, after 45 years, the evidence trail, which has taken us to so many fascinating parts of Australia and the wider world, has led us to the point where we have to acknowledge that, while the creatures encountered by our many informants are undoubtedly real, they are not "real" in the usual sense of the word. Decidedly uncanny, they may forever remain beyond human understanding.

Mind you, if a yowie suddenly takes it into its head to walk into a country pub, sit down and pose for selfies with the locals, we'll be more than happy to change our minds.

CHAPTER ONE
THE COLONIAL ERA, 1788–1900

Case 1. Botany Bay, NSW, 1789

Unlike most items in our collection, this one requires quite a bit of explanatory text.

It is an obvious hoax, but we've included it for two reasons. Firstly, because it is the earliest written reference to something resembling the hairy giants of Indigenous legend, and secondly out of fairness to sceptics, who, noting its similarity to ancient European "wild man" traditions, might like to suggest such traditions influenced British settlers' interpretations of Indigenous tales.

The story comes from a handbill circulated in Britain in about 1790. It concerns a nine-foot-tall hairy wild man supposedly captured at Botany Bay in 1789 and taken to England. In reality, if such a remarkable creature had ever set foot in "the Old Dart", it would have been exhibited all over the country, comprehensively poked and probed, and eventually stuffed. The hapless critter would now be staring out at us from a glass case in the British Museum.

Anyway, this, for what it is worth, is the handbill's (lightly edited) text:

> "THERE have been various reports concerning this most surprising wild man, or huge savage GIANT, that was brought from Botany Bay to England … Thousands have seen him in Plymouth, where he was landed alive and in good health.
>
> This surprising monstrous giant was taken by a crew of English sailors when they went on shore to furnish themselves with fresh water at Botany Bay … they beheld at a distance three of the most surprising tallest and biggest looking naked men that have been seen in the memory of this age … which … caused them to … board the ship for the safety of their lives, leaving the casks of water and a quantity of good old rum …
>
> When the three savages got to the seaside they stared at the ship … with wonder and admiration, one of them having got the cag of rum, he tasted, spit it out and shook his head, another did the same, but the third drank plentifully, and began to jump about in a frightful wild manner, shouting and making a hideous noise. The other giants went off and left this one … who drank to such excess that he dropped on the ground and lay as if dead. The sailors went on shore … bound him fast with ropes and … got him on board the ship, where they secured him with iron chains. He came in the ship Rover, Capt. Lee, to England from Botany Bay, and landed at Plymouth, November 29, 1789.
>
> He is … not so savage in temper as might be expected. He is 9 feet 7 inches high, 4 feet 10 inches broad, a remarkable large head, broad face, frightful eyes, a broad nose and thick lips … very broad teeth, heavy eye-brows, hair stronger than a horse's mane, a long beard strong as black wire, body and limbs covered with strong black hair, the nails of his fingers and toes may be properly called

talons, crookt like a hawk's bill, and as hard as horn. In short, he is viewed with admiration and astonishment on account of his huge size.

A defcription of a wonderful large WILD MAN or monftrouf GIANT, BROUGHT FROM BOTANY BAY, handbill circa 1790. Collection of the State Library of New South Wales.

The "Monstrous Giant" of Botany Bay – a woodcut from the handbill

The Botany Bay "wild man" – again

Some years after we became aware of that unlikely tale, we were alerted to the existence of a second illustration that was probably inspired by the same tall story. It first appeared in a thirteen-page booklet titled *A tour through Apollo Gardens, in Gawsworth, near Macclesfield, Cheshire*, printed in 1802 by J. Dean, of Congleton, Cheshire.

In the entirely invented, dream-like "tour", the narrator relates how, after viewing several other marvels in the gardens, he is confronted by the "Wild Man of Botany Bay", who, though chained, rushes at him brandishing a club:

> His body was all covered with hair at least two inches long; and over his eyes it was somewhat longer … His beard was monstrously long and black, as was the hair on his head, which hung clotted down his back … I supposed him to be about six feet high … before his den lay a heap of bones …

Researcher Nat Williams unearthed the story while perusing *Portraits of the Famous and Infamous: Australia, New Zealand and the Pacific, 1492–1970*, a massive collection of images that was gifted by Rex Nan Kivell to the National Library of Australia some years ago.

"The Hairy Wild Man from Botany Bay" c. 1802. (Rex Nan Kivell Collection, National Library of Australia)

The Hairy Wild Man from Botany Bay

The "wild men" of Europe

There are, of course, many reasons to doubt the authenticity of the Botany Bay wild man story. Folklorists will have noticed that, particularly as depicted in the second illustration, the creature looks remarkably similar to the "wild men" that frequently appeared in the art and literature of medieval Europe. In the Middle Ages, it was widely believed that such huge, hair-covered, man-like monsters roamed the dense forests that still covered vast swathes of the continent. In Britain they were often referred to as "woodwoses" or "woodhouses", corruptions of the Anglo-Saxon term wudewasa, meaning "wild man of the woods".

Some Britons, therefore, might have been prepared to believe that something resembling the legendary European wild man existed in the vast unknown of New South Wales.

But although the "monstrous giant" story is undoubtedly a complete fabrication, we have always been amused by the fact that the huge, yowie-like creature was supposedly captured in 1789 at Botany Bay – the very place where, just a year earlier, British colonists, led by Captain Arthur Phillip, first set foot in Australia.

Just one kilometre from the site of Captain Phillip's landing there is an unusually-shaped jumble of scrub-covered boulders known to residents of the La Perouse Aboriginal Reserve as "Hairy Man Rocks." Even today, Aboriginal children are warned that if they venture there after dark, the *doolagarl* might grab them.

So it seems the Hairy Man was right under the noses of the British from "Day One."

It is possible, therefore, that the 1789 "Wild Man" story, bogus as it is, might have been inspired, in part, by Aboriginal stories about yowies in the area. We find it interesting, also, that Botany Bay is only eight kilometres north of Port Hacking, where several encounters with giant, hairy "yahoos" were reported in the 1850s and 1860s. (See Cases 24 and 25). The area has also produced two very credible modern-era yowie reports. (*The Yowie*, pp. 26-27, 255 and 284-285).

Case 2. 1820s. Hunter Valley, NSW. Aboriginal lore

Although it is difficult to pinpoint the earliest actual sighting report by a non-Aboriginal, it is certain that Europeans became aware of Aboriginal yowie lore in the first couple of decades of the 19th century, if not earlier. Captain Peter Cunningham, a Royal Navy surgeon who farmed in the Hunter Valley near Mt Sugarloaf in the 1820s, may have been the first European to write about the phenomenon.

In his memoir, *Two Years in New South Wales*, he mentioned that the local Awabakal people believed in a fearsome, man-eating creature called *puttikan*, which resembled a tall man with a hairy body and long mane. Its feet were reversed to confuse trackers and its skin was so tough that spears could not pierce it. It roamed by night and devoured children but was afraid of fire.

Cunningham believed fear of *puttikan* was the reason his Aboriginal neighbours never travelled at night and always slept close to a fire. On still summer evenings, he heard strange cries attributed to the hairy giant echoing through the mountains, so, despite the bizarre reversed feet detail, he seemed inclined to give the legend some credence.

As we will see, colonial-era Aborigines often asserted that the mysterious creature's feet were reversed on its ankles – a truly bizarre and implausible belief. It's worth noting that "reversed foot syndrome" is a classic folklore motif that crops up all over the world in legends about various demons – and is mentioned in several other ape-man legends.

In Nepal, for instance, it is a standard feature of yeti lore, and Malaysian aborigines, the Orang Asli, believe the same thing about the hairy, super-elusive *orang mawas*. Even on remote Andros Island, in the Bahamas, locals say the feet of a fearsome ape, the *yay-ho*, point backwards.

In any case, we note that Cunningham's testimony is nicely corroborated by that of Rev. L. E. Threlkeld, who oversaw a mission to the Aborigines in the same region during the 1820s and early 1830s. (Case 3, below)

Cunningham, Peter, *Two Years in New South Wales: A Series of Letters, comprising Sketches of the Actual State of Society in that Colony; of its Peculiar Advantages to Emigrants; of its Topography, Natural History, Etc., Etc.*, Henry Colburn, New Burlington Street, London, 1827.

Case 3. 1820s or early 1830s. Newcastle district, NSW

According to the *Newcastle Morning Herald* of 2 October 1912, Rev. L.E. Threlkeld of the London Missionary Society wrote the following from Newcastle in December 1825 concerning his mission to the Aborigines:

> "The aborigine who assists me in obtaining a knowledge of the native language informs me that there is a being in the Sugarloaf Mountain, resembling a man, but taller in stature, with arms, legs, face, and hair, the latter very long on the head; but the feet are placed contrary to the head, being behind; and the body is hairy, like that of an animal.
>
> The flesh of this being is so hard in all parts that it is impenetrable except just between the legs, where a spear may penetrate, but in no other part. He is a fierce being, devouring men, and often pursues the aborigines in the mountains. There are families of this being; but not many of the species.
>
> The cry of these beings is often heard uttering the sound of "Perrolorl-o", dwelling very long on the 'O'. The sound is heard in summer time. I inquired whether any Europeans had ever seen this Achilles of the aborigines, and the reply was given that a soldier of the 46th Regiment had heard him one night when he was hunting with the aborigines.

Rev. L. E. Threlkeld

Threlkeld goes on to say that the fearsome creature was known as *Yarho Patiogarang*. The information gathered by Threlkeld, of course, nicely corroborates that recorded by Captain Peter Cunningham.

Case 4. c. 1820s. Port Macquarie hinterland, NSW

Because several newly discovered rivers seemed to run westward, it seemed plausible, during the early years of European settlement, that there could be a huge lake, even a small sea, deep in the heart of the unexplored continent.

A Sydney *Colonist* article concerned mainly with that idea also mentioned stories of strange animals that a missionary heard from Aborigines to the west of Port Macquarie "several years" prior to 1836:

> ... the natives drew upon the sand the representation of an animal which came out of the water and was destructive to them, and also another animal like ... the ourang-outang, which they say carried away their [women].

The Colonist, Sydney, 4 Aug 1836

Case 5. c.1820s-1840s. NSW south coast. Aboriginal lore

Although colonial-era Aborigines, like their modern-day descendants, insisted the Hairy Men (aka *doolagarls*, etc.) were real, flesh and blood creatures, it is undeniable that there were folkloric and supernatural elements in many, if not most, of their stories, as recorded by interested white pioneers.

In 1897, in an article titled "Aboriginal Superstition", Robert Brothers recalled hearing two Hairy Man stories at Tabourie Bridge, between Jervis Bay and Batemans Bay. His informant was an 87-year-old Aboriginal woman. She referred to the creatures as *thooligals* and said she'd seen two of them several decades earlier.

Brothers quoted her as saying "We went down to the beach one day and my sister said 'Hush!' like that. We looked. There, upon the rocks where the waves were breaking, it sat. A crown of flowers was on its head, and its hair was like the sun shining on water. It did not see us, for it was looking away over the sea. We crept away, and when we got far enough, we began to run. My side was aching when we heard it coming behind us. That was the first time I saw it."

The second incident occurred as she and her husband, while travelling, were camped for the night. Waking after a couple of hours, she saw a *thooligal* glaring at her across the fire. Her husband, who was a *bangal* (shaman), advanced toward the uninvited visitor, buried his tomahawk blade downwards in the earth and recited an incantation.

> "I hid my face in the rug", the old lady recalled, "then [looked out, and] saw its eyes like coals, with hell in them." The creature eventually yielded to the exorcism and ran off with a fearful yell.

Evening News (Sydney), 27 Feb 1897

Case 6. c.1820s – 1831. Joseph Bonaparte Gulf (?) WA/NT border

In 1831, a convict who'd been at large for many years surrendered to Major McPherson, the Commandant at Bathurst, and told of making "a most important discovery … to the Northward of the colony".

He claimed to have followed "a noble river" all the way from "Northward of the Liverpool Plains" to "the gulf of Van Diemen … nearly opposite of Timor" where he "fell in with several tribes of natives, armed with bows and arrows" and seafarers "presumed to be parties of Malays … who come over to procure sandal wood and beche-le-mer."

He also told of encountering "numbers of Hippopotamuses [sic] and Ourang Outangs – animals of whose existence in New Holland we have never before heard even a surmise."

It wasn't unusual for runaway convicts, on surrendering, to claim (in the hope of leniency) discoveries that might intrigue the authorities. While some such tales were true, this one is definitely – in large part, at least – bogus.

There is no "noble river" flowing from the Liverpool Plains to anywhere remotely approaching Australia's northern coast, and, although bows and arrows were used on the Torres Strait islands (well to the east of Joseph Bonaparte Gulf), there is virtually zero evidence for their use on the Australian mainland.

All that, of course, makes us wonder if there's any truth in the unnamed convict's claims of encountering hippos and orang outangs.

He might conceivably have travelled far enough north to have seen vaguely hippo-like dugongs, and he might – just possibly – have also encountered the rather orang outang-like yowie during his travels.

Probably not. But as this unlikely story is the earliest in our files that involves a claimed sighting of hairy ape-like beings by a non-Aboriginal, we are reluctant to simply "bin" it.

Sydney Gazette, 1 Nov 1831

Case 7. 1830s. Hunter Valley, NSW. Aboriginal lore

In 1847 Alexander Harris wrote of his experiences in the Hunter region some years earlier:

> "The blacks were occasionally, but not often, troublesome. The stories they used to tell us about the brush thereabouts being haunted by a great tall animal like a man with his feet turned backwards, of much greater, however, than the human stature, and covered in hair, and perpetually making a frightful noise

as he wandered alone, made me sometimes doubt whether they were themselves really terrified, or were merely endeavouring to scare us away; but I very strongly incline to the latter opinion.

Alexander Harris, *An Emigrant Mechanic, Settlers and Convicts, or Recollections of Sixteen Years Labour in the Australian Backwoods*, London 1847. Reprinted, Melbourne University Press, 1969.

Alexander Harris

Case 8. 1830s or '40s. The Gwydir district, near Moree, northern NSW

In our previous book, *The Yowie*, we suggested that this incident occurred during the 1870s. Now, having obtained the entire document in which it is recorded, we know it is much more likely to have happened thirty or more years earlier.

Charles Naseby first went to the Gwydir in 1833, when Aborigines were spearing a lot of cattle, when "white men ... began to use their guns freely [and when] the natives ... were killing all white men they could find." Although many settlers, such as Naseby himself, refused to take part, and though some Aborigines were protected from the worst of it, the shameful episode that he referred to as "that war of extermination" continued into the 1840s.

As the worst of the violence seems to have been over by the 1850s, we feel that the incident in question, involving as it does two men who "yarded their sheep at sundown for fear of the blacks", and thought nothing of firing directly at "an object like a blackfellow", probably occurred during the 1830s or '40s.

"I have heard a Shepherd tell a story about a *dibble dibble*, [Naseby wrote] but as the incident does not fall within my own experiences I do not vouch for its truth.

Geo. Long and his mate had yarded their sheep at sundown for fear of the blacks, and were sitting down to their supper, when the dogs came running into the hut in great alarm; the shepherds drove them out, but presently they returned, evidently frightened by something outside.

Long rose and peeping cautiously out the door he saw under a tree at a distance, an object like a blackfellow, but considerably larger. Whispering to his companion to bring him the gun, he raised it to shoulder, and aiming at the breast of the blackfellow he fired.

At once the object began to move, and with huge strides fled away across the river, making a clattering noise on the shingle, as if with the hoof of a horse. Unable to satisfy themselves what it was, they next morning obtained the assistance of a blackfellow, and proceeded to examine the track.

They discovered several footprints, and as soon as the black saw the first of them he exhibited every symptom of terror, and muttered *dibble dibble*. They said, "Nonsense, *dibble dibble* does not care for white men." "No," he said, "but *dibble dibble* has come to look for blackfellows."

The marks were like those of an emu's feet, but there was one long claw which penetrated several inches into the ground.

Naseby, C., "*Stories about the Kamilaroi Tribe*" in Fraser, John, *The Aborigines of Australia*, Tucker, Gillies and Thompson Printers, "Mercury" Office, High Street, Maitland, 1882, p. 10.

The following three items were taken from the *Australian and New Zealand Monthly Magazine*, 1842, pp. 92-96.

Case 9. c. 1836. Onkaparinga River, 30 km south of Adelaide, SA

"The natives of Australia … believe in the imaginary existence of a class which, in the singular number, they call Yahoo, or, when they wish to be anglified, *Devil-Devil.*

This being they describe as resembling a man, of nearly the same height, but more slender, with long white straight hair hanging down from the head over the features, so as almost entirely to conceal them; the arms as extraordinarily long, furnished at the extremities with great talons, and the feet turned backwards, so that, on flying from man, the imprint of the foot appears as if the being had travelled in the opposite direction.

Altogether they describe it as a hideous monster, of an unearthly character and ape-like appearance.

The dread of this spectre deters them from venturing abroad after sunset, unless in numbers, and having fire with them, which they conceive intimidates the fiend … one of the many evil endowments which the natives attribute as belonging to this fanciful creature, that of carrying off children and females, no traces of whom are afterwards found, appears to be the most prominent and dreaded.

They also affirm, but with less apprehension, that it occasionally attacks men when single-handed and in the dark; but they do not consider it as equal to one of themselves in an encounter by day …

The supernatural agency of the Yahoo is very limited, and confined chiefly to such occurrences as the following: Appearing suddenly and unaccountably, pouncing upon and destroying the defenceless, and maintaining certain baleful influences over human life …

Reverting to the subject, the following is a story related by Mungaroke, a native of a tribe near Adelaide, of a Yahoo.

A gentleman, who [travelled] from Encounter Bay to Adelaide at the first settling of the colony [1836] took with him some natives for guides … and passed the Onkaparinga, where [the party] rested for a short period among the sand hills at the mouth of the river. [Every member of the party was listless and inactive from fatigue] … with the exception of Koteragee, upon being questioned as to the cause of his restlessness, with evident circumspection informed them that he had once had a very serious adventure upon the [same] spot with a yahoo …

He, Koteragee, accompanied by two males and one female, were returning from a visit to the tribe resident near Adelaide, when they arrived at the banks of the river Onkaparinga, where Munchee Munchee … proposed that they should halt and … cook their lizards, etc., and refresh themselves … the males proceeded accordingly to the bottom of the dell for water, when, to their great surprise and dismay, they espied something lying beside the spring, apparently in a sound sleep, which they immediately recognised to be a Yahoo …

After a little deliberation the trio wound up their courage, and determined at last to rid the world of one monster. Accordingly, stepping up to the object of their hate and fear as noiselessly as shadows, they simultaneously discharged a shower of blows on the head of their intended victim, notwithstanding which, the semi-incarnate being sprang upon its feet and attempted resolutely to escape; but finding itself hemmed … it darted at Bidjoke (one of the trio) with its fiendish talons, who, however, eluded the monster with a backward leap; upon which Munchee Munchee flew at the demon, and with a terrible stroke to the temple brought it to the ground; and then, with Bidjoke, fell upon and held the Yahoo fast, while Koteragee drove his waddy through the skull from ear to ear.

Being determined to leave no chance of reanimations they threw the carcase into the river … [but] … The body, although apparently motionless, by some unaccountable means floated to the opposite side … and to their utter consternation and amazement started at once into life and bounded off into the furze that grew on the opposite bank.

This story, which was subsequently confirmed by Bidjoke, Munchee Munchee and Koteragee, evidently cost poor Mungaroke a great deal of pain; for while he doled it out, the furtive glances that he occasionally cast over his shoulder too plainly told the apprehension he felt of the Yahoo's reappearance, although the native superstition teaches that it will not approach the presence of the white man.

The preceding narrative, although no doubt an exaggerated statement, suggests the possibility that this object of terror … might possibly have been either a man or an animal of the lower grade. It would certainly by no means appear improbable that the unfortunate object … might have been some unfortunate castaway sailor, runaway convict or sealer from Kangaroo Island unshaven and unshorn for weeks or years. In favour of this suggestion, it should be known that Old Con, a native of Adelaide, tells his white friends that when Captain Sturt first entered Lake Alexandrina, he and his party were taken by the natives for Yahoos.

On the other hand, a contested point has long existed among Australian naturalists whether or not such an animal as the Yahoo existed, one party contending that it does, and that from its scarceness, slyness, and solitary habits, man has not succeeded in obtaining a specimen, and that it is most likely of the monkey tribe.

In favour of this assertion they bring the united testimony of the natives and settlers to support their opinion; from the latter … two instances of its appearance have been obtained …

Case 10. c. 1830s. Near Sydney, NSW

“… a few years ago a Yahoo exhibited itself in the government domains of Sydney before several persons, but quickly effected its escape upon their approach.

Case 11. c. 1830s. Kangaroo Island, SA

“… a man named Thompson, who for many years lived upon Kangaroo Island; he being questioned upon the subject, gave the following anecdote.

Being out one day in a boat with two or three others, he landed in a creek (which he still points out) and to his astonishment observed a being corresponding to the description given by the natives of the Yahoo, sitting upon a point

> of a high isolated rock, viewing him very unconcernedly. Levelling his musket he fired, when the being, shrieking wildly, disappeared among the contiguous recesses.
>
> The appearance of this creature, and the peals of agony that burst on his ear after firing, have made such an impression upon this man's mind that at times both are conjured into existence by the bare recital, and not infrequently by the moaning breeze.

"Superstitions of the Australian Aborigines: The Yahoo", *Australian and New Zealand Monthly Magazine*, 1842, pp. 92-96.

This incident is the earliest we know of wherein a non-Aboriginal yowie witness is identified by name. It is also the first of a handful of reports (e.g., Cases 21 and 204) that tell of yowies being sighted on offshore islands.

Kangaroo Island is 14 kilometres from the mainland – quite a long swim, one imagines, for any land animal.

Case 12. c. 1839-1844. NSW and Tasmania

Louisa Anne Meredith (1812-1895) arrived in Australia with her husband Charles in 1839 and after a year at Homebush, NSW, moved to Tasmania where she lived for the rest of her long life. A prolific writer and gifted artist, she published books of poetry, two novels, several books of Australian natural history, and two interesting accounts of her early years in the colonies: *Notes and Sketches of New South Wales during a Residence in the Colony from 1839 to 1844* (1844) and *My Home in Tasmania* (1852).

Although she was considered, in her day, to be a progressive thinker (before leaving Britain she championed the Chartists' cause) her comments about Indigenous Australians reveal she was still very much a product of her time.

> [The Aborigines] have an evil spirit, which causes them great terror, whom they call "Yahoo" of "Devil-devil": he lives in the tops of the steepest and rockiest mountains, which are totally inaccessible to all human beings, and comes down at night to seize and run away with men, women, or children, whom he eats up, children being his favourite food; and this superstition is used doubtless as a cloak to many a horrid and revolting crime committed by the wretched and unnatural mothers, who nearly always, when the infants disappear, say "Yahoo" took them.

They never can tell which way he goes by his tracks, because he has the power of turning his feet in any direction he pleases, but usually wears them heels first, or, as they express it, "Mundoey that-a-way, cobbra *that*-a-way" (feet going one way, and head or face pointing the other).

The name Devil-devil is of course borrowed from our vocabulary, and the doubling of the phrase denotes how terrible or intense a devil he is; that of Yahoo, being used to express a bad spirit, or "Bugaboo", was common also with the aborigines of Van Diemen's Land, and is as likely to be a coincidence with, as a loan from, Dean Swift.

Louisa Anne Meredith

Mrs. Charles Meredith, *Notes and Sketches of New South Wales during a Residence in the Colony from 1839 to 1844* (London, 1844) p. 95.

Case 13. c. 1840. Moruya, NSW. Aboriginal lore

While on a US Navy voyage of exploration, Horatio Hale spoke with an Aborigine at the present site of Moruya:

"At the Muruya River the devil is called Tulugal. He is described as a black man of great stature, grizzled with age, who has very long legs, so that he soon overtakes a man, but very short arms, which brings the contest nearer an equality.

This goblin has a wife who is much like himself but still more feared, being of a cruel disposition, with a cannibal appetite, especially for young children.

Horatio Hale, *United States Exploring Expedition, During the Years 1838, 1839, 1840, 1841, 1842, Under the Command of Charles Wilkes, U.S.N. Ethnography and Philology.* Philadelphia: Lea and Blanchard, 1846.

Horatio Hale

Case 14. April 1843. Mangrove Creek, 12 km west of Gosford, NSW

From the *Sydney Morning Herald*, 12 July 1843:

> "We are credibly informed that several parties have been unexpectedly alarmed by the sudden appearance … in the neighbourhood of Mangrove … of a monster, which they designate a Yaa-hoo (from the noise he makes), who has come upon persons when encamping at night and to others in the vicinity of their houses, he is described as being nearly seven feet in height, with feet apparently like human feet, only of enormous size, with his visage turned backwards, and his body covered with apparently a down; he appears to be harmless, never having been known to commit violence, and always utters his uncouth Yaa-hoo.
>
> There are parties who persist in stating that he was close to their house in Mangrove, and others who declare to having met him on the road to Wicketty-Wees! We merely give the facts as stated to us … and which upon enquiry, we find firmly persisted in.

The assertion that the creature's head was reversed on its neck is rather interesting in light of the belief among some Aborigines that its *feet* were reversed.

This is also the first report we know of in which colonists say that the creatures were referred to as "Yahoos" because of their vocalisations.

A week later, on 19 July, the *Herald* ran a much more interesting follow-up article:

> "We … add that the oral testimony [of] eye-witnesses whose veracity we have no reason to doubt, is fully borne out by their serious and impressive description … In the desire to elicit enquiry we subjoin the best authenticated account we have collected.
>
> About three months since, two young people (brother and sister) being left at home in Mangrove Creek about mid-day, hearing a strange noise, and the dogs barking violently, went out, and upon looking in the direction of a myrtle-scrub opposite, to which their attention was directed to the dogs, they observed a dark hairy figure, which, from its height above the scrub, they conjectured to be nearly eight feet. Against this figure the rage of the dogs was directed.
>
> Being alarmed, [they] ran for a neighbour, who, upon arrival was equally frightened at the strange visage and uncouth sounds of the monster. Shortly after, the dogs made a rush, and with the figure went off in an opposite direction; one of the dogs, we should observe, remained away above half an hour, and returned quite exhausted.

Up to that point, the story reads exactly like a yowie report of the modern era. Thereafter, however, it takes a rather folkloric turn:

> Upon the proprietor of the farm coming home, and learning what had occurred, he again put the dogs into the scrub, and upon entering himself, he could find no other traces of a living being, than the fresh track of apparently very small hoofs in the direction pointed out.
>
> We have merely recorded the oral statements of professed eye-witnesses, whom we feel satisfied are confirmed in the reality of what they witnessed, and the correctness of their description, however sceptical we ourselves may feel disposed to be upon certain particulars, especially the hoofs, which savour strongly of Pan, or the Evil One.

Sydney Morning Herald, 12 and 19 Jul 1843.

In September 1918, another yowie was seen at Mangrove Creek. That one left behind some more plausibly-shaped, 18-inch-long tracks. (Case 137)

Case 15. 1840s. Hunter River, NSW. Aboriginal lore

In the following article, the Hairy Man phenomenon has become confused with that of the legendary river monster, the bunyip.

> "At the Hunter's River the reports of the natives would lead us to classify it with the carnivorous species. In this locality it is called *Yaa hoo*, and is described as having much resemblance in form to the human figure, but with frightful features – the feet like those of a man, but reversed or turned backwards … In the immediate neighbourhood of the river the animal is called *Wowee Wowee*, and the blacks picture its haunts as and habits as purely aquatic.

Sydney Morning Herald, 21 Jan 1847.

Case 16. 1840s. South Australia. Aboriginal lore

> "The natives here have a tradition that a big black fellow, far higher than the ordinary size, walks about during the night, his object being to destroy good blackfellows and their children – the latter articles being his favourite diet; and they will sometimes show a foot-print, in size about three times as large as an ordinary foot, and in shape resembling the print of a man's step. He is said to walk about principally during the night, for which reason they never stir out at that time. I asked one of the men why he did not kill this creature. Upon which he replied with much earnestness, "Oh! me plenty run away- me too much frightened!" They give this being the name of Noocoonah.

"A Squatter", *Reminiscences of A sojourn in South Australia*, Kent and Richards, London, and George Phillip and Son, Liverpool, 1849.

Interestingly, this story suggests that while British colonists had not actually sighted the *noocoonah*, some of them had examined its tracks.

Case 17. 1847. Near Cape Otway, VIC

Published in the *Sydney Chronicle* on 28 August 1847 under the headline "THE BUNYIP OUT-BUNYIPPED", this item is a clumsily-written tall story. Although the *Chronicle* apparently thought it might amuse its colonial-era readers, it now seems decidedly lame. As it runs for an excruciating five pages, we present only a brief summary.

While riding through the scrub, the unnamed writer paused beside a freshwater lake, where he was seized from above by "a cold bristly hand" and hoisted out into the branches of a large tree. His captor was "a huge animal nearly corresponding to the ourang of the Eastern Archipelago," which proceeded to caper around in a slap-stick manner.

After some hours, the eight-foot-tall monster stripped him of his clothes. As it struggled to don his trousers, the stolen hat fell over its eyes and the man lashed out, sending it "roaring into the water below," and made his escape.

Although this irritating yarn is clearly not meant to be taken seriously, it does feature an animal somewhat like the Hairy Men described in other, apparently serious, reports. While the writer doesn't employ the term "yahoo", and while his monster (unlike those in almost all of our reports) is entirely arboreal, it's possible he based his tale on Aboriginal lore or on rumours of real encounters with yowies.

While the story is almost certainly bogus, we can't help noticing that the area in question has produced two excellent modern-era yowie reports – one involving a wildlife ranger. (Case 242 and *The Yowie*, p. 254.)

Case 18. c. 1847. Junction of the Yass and Murrumbidgee Rivers, NSW

A Ngunnawal/Kamberri elder, Harry Williams (c.1837-1921) told his friend George Webb that as a child he saw a large group of Aborigines kill a Hairy Man. The incident occurred close to the present site of Burrinjuck Dam.

In a letter to the *Queanbeyan Age*, Mr. Webb wrote:

> Having interviewed Harry Williams re the animal he saw the blacks kill, he pointed out to Mr. J. McDonald and myself the height he (Williams) was at the time, and we considered he would be from 10 to 12 years old.
>
> The locality where the blacks killed it was below the junction of the Yass River with the Murrumbidgee. The animal got into some cliffs of rocks [sic] and the blacks got torches to find out where it was hidden and then killed it with their nullah nullahs. There was a great many blacks at the killing, and he saw two

dragging it down the hill by its legs. It was like a black man, but covered with grey hair.

Queanbeyan Age, 7 Aug 1903.

In 1987, just ten kilometres south-east of the place where Mr Williams saw the Hairy Man killed, James Basham of Cootamundra came within a few feet of a similar creature in the middle of Taemas Bridge, over the Murrumbidgee River. (*The Yowie*, p. 258).

Harry Williams

Case 19. c. 1847. Near Western Port Bay, VIC. Aboriginal and colonist lore

"THE WILD MAN OF THE AUSTRALIAN WOODS

A creature described by the natives as something very similar to an ourang-outang is supposed by many colonists to exist in the mountain ranges at the back of Western Port, but their ideas of it are mixed up with such a superstitious dread as to induce many to consider it only in the light of an imaginary being, created by their own fears, or by interested parties amongst themselves; but the fact of some strange and peculiar tracks having been noticed in the ranges, recorded in the Port Philip papers at the time … and many other circumstances, seem to indicate that there is some animal resident there which has not yet been seen by a white man …

An account of this animal was given me by Worrouge-toulon, a native of the Woe-worong tribe, in nearly the following manner: "He is as big as a man and shaped like him in every respect, and is covered with stiff bristly hair, excepting about the face, which is like an old man's, full of wrinkles; he has long toes and fingers, and piles up stones to protect him from the wind or rain, and usually walks about with a stick, and climbs trees with great facility; the whole of his body is hard and sinewy, like wood to the touch."

Worrongbe [sic] also told me that many years ago some of these creatures attacked a camp of natives in the mountains, since which period they have had a dread of moving about there after sunset. The only person now alive who killed one, he told me, was Carbora, the great doctor, who succeeded in hitting one in the eye with his tomahawk. On no other part of his body was he able to make the least impression.

On one occasion, when pheasant shooting about three days journey in the mountains, in company with two natives and a white man, we constructed a bark hut, and had retired to repose, when shortly afterwards, I was startled by a most peculiar cry, very different from any of the other noises which are heard from the wild animals inhabiting these ranges, I should have previously mentioned, that the blacks, after the fatigues of the day, had very soon fallen asleep; but, on the noise … they both started up and seized their guns, with the utmost horror depicted on their countenances. Not a word escaped them, and the mysterious sounds still echoed amongst the hills.

On my asking one, in rather a loud voice, what he was frightened at, he desired me not to speak loud; that the shouts which had aroused them proceeded from a bundyilcarno, or devil, which is the name they have given this thing. The noise shortly died away in the distance, and I once more endeavoured to sleep. Neither of my natives would lie down for the night, and as soon as day dawned, they insisted on leaving the scene of this strange occurrence and going to some distant part.

The Perth Gazette and Independent Journal of Politics and News, 26 Feb 1848.

A baboon-like animal was reportedly seen by colonial officials on Phillip Island in Western Port Bay two years later. (Case 21)

Case 20. c. 1848. Near Cudgegong, NSW

“About thirty years ago a shepherd in W. Sutton's employ averred he had seen a hairy man in the scrub north of Cunningham's Creek … he persisted to the day he died that it walked upright and was covered with hair, and the dogs that hunted everything else ran back from this frightened with their tails between their legs.

Significantly, perhaps, an elevation just west of Cunningham's Creek has been known, since at least the 1860s, as Monkey Hill.

Lismore Northern Star, 17 May 1878.

Case 21. 1849. Phillip Island, Western Port Bay, VIC

"We are informed by Mr Edwards, the managing clerk, at the office of Messrs Moor and Chambers, that during his late trip … to capture the runaway Hovenden, that while on … Phillip Island, he and his party were astonished at observing an animal sitting upon a bank in a lake. The animal is described as being from six to seven feet long, and in general appearance half man, half baboon.

Five shots were fired … upon the first shot whistling past him, he appeared somewhat surprised, and shook his head … at the second, he grinned fiercely and showed an uninviting set of teeth; at the third he backed towards the water; the fourth was answered by a half growl, half shout, which made the "welkin ring", and the fifth … was replied to, by a spring into the air, and a contemptuous fling out of the hind legs and a final disappearance in the placid waters of the lake. A somewhat long neck, feathered like the emu, was the peculiar characteristic of the animal.

The Argus, 25 Oct 1849; *Melbourne Morning Herald*, 29 Oct 1849; "Bell's Life", in *Sydney and Sporting Reviewer*, 10 Nov 1849.

Despite the "long, feathered neck" and the journalist's jokey tone, it is possible this odd story was based on a real incident. In a long postscript, full of ponderous colonial-era humour, the editor of the *Argus* admits that he invented the detail of the creature's leaping into the air and flicking its legs. As the story came to the *Argus* via the *Melbourne Morning Herald*, it is possible the other inconvenient detail – the long, feathery neck – was also added somewhere along the way.

As we have seen, (Case 19) this story is not the first to suggest yowies were haunting the Western Port Bay area in the 1840s.

It is also not the first report of a yowie on an offshore island – one of the critters was seen on Kangaroo Island, 14 kilometres off the South Australian coast, in the 1830s (Case 11). In this case, however, the creature had only to swim 800 metres to reach the Victorian mainland.

Case 22. 1851. Oakey Creek, near Mudgee, NSW

This tantalising story is one of only two in our files that concern the supposed discovery of large ape-like skulls.

> Part of the skull of some unknown animal was last week dug out of the bank of the river, about a quarter of a mile below Oakey Creek. It was found thirty feet below the surface of the earth, along with some other bones belonging to the same animal, apparently. The person who saw it informed me that the upper and lower jaws, as well as the teeth were all perfect, and that they were unlike any other skull bones that he had ever seen, and appeared more than anything else to resemble the head of a large monkey, the teeth being somewhat like those of a human being, but rather larger. I shall endeavour to see it myself, when I shall perhaps be able to give a better account of it.

"News from the Gold Fields", *Sydney Morning Herald*, 2 Oct 1851.

The Mudgee – Pyramul area was something of a yowie hot spot in colonial days – and the creatures continued to haunt the area well into the modern era. (e.g., Cases 42 and 127)

Case 23. 1851. Lake Torrens, SA

Ernest Favenc

> In 1851, two squatters in search of a run, Messrs. Oakden and Hulkes, pushed out to the western side of Lake Torrens, and according to their account found a most favourable land. They discovered a lake of fresh water, surrounded with good country; and the natives told them of other lakes to the north-west; also introducing descriptions of strange animals, whose appearance could only have been equalled by that of the jimbra, or apes, of Western Australia, which ruthless animals, according to blackfellow's legend, devoured the survivors of Leichardt's party, as they struggled into the confines of that colony. (See Case 26)

Ernest Favenc, *The History of Australian Exploration from 1788 to 1888*, Turner and Henderson, Sydney, 1888, p.188.

It is worth noting that in 1995, 144 years after Oakden and Hulkes were warned about the ape-like beings, 11-year-old Jarrod Nicholson and some friends encountered a similar creature near Roxby Downs, 35 kilometres west of Lake Torrens. (Case 221)

Case 24. 1856. Port Hacking, NSW

Captain William Collin

In his autobiography *Life and Adventures (of an Essexman)* Captain William Collin said that he and a friend named Massie received dire warnings about a giant, hairy yahoo while camped on the southern shore of Port Hacking in 1856. The story is of considerable interest because it is our only colonial-era case in which a protagonist specifically mentions the fictional yahoos of *Gulliver's Travels*.

In 1856 there was only one white family, the Gogerlys, living at Port Hacking [at a spot still known as Gogerlys Point]. The clan's patriarch, Charles Gogerly, was, as Captain Collin put it, "a curious old gentleman".

Collin and his mate were at the bay to gather seashells for lime making: "We did very well, selling the shell to schooners and ketches which carried it to Sydney. For some reason or other Gogerly did not like us in his vicinity, and ingeniously worked up a scheme which he vainly thought would frighten us away. One afternoon Gogerly sent down two of his boys in an old log canoe, to tell us that [he] had seen a Yahoo, or wild man of the woods; it was about 12 feet high, they said, and carrying a staff 20 feet long [and that] we were not safe … as it was seen close to our tents." Later, old Mr Gogerly claimed that "on hearing a noise … he used his spyglass … and scanned the shore, till his eyes rested on the monster, which he declared was looking at my mate and myself, as we gathered shells …".

Massie took the warning seriously: "My mate … who was a great reader, and who had no doubt dived at one time or another into 'Gulliver's Travels', said he knew such things as yahoos existed; and as there were a number of deep gullies about, this was no doubt a likely place for them." Although Collin insists, in his memoir, that he always thought Gogerly's story was an invention, it is clear from his actions that he was not entirely convinced of that:

"'If we could only trap this Yahoo', I told my mate, 'we should not need to trouble any further about gathering shell'. We loaded our guns, and ... certainly did find some remarkable tracks, which had not been made by a human being. What they were I had no idea. Massie … would not sleep on the shore, feeling safer in the boat. I eventually lay down with my gun alongside me. The place was infested with dingoes … which made noise enough for half a dozen Yahoos. In fact, I made up my mind one night that the creature was upon us, and fired, only with the result of setting a dingo yelling most pitifully. We shifted camp ... and were not further disturbed by either wild dogs ... or the mythical Yahoos."

Captain William Collin, *Life and Adventures (of an Essex-man)*, H J Diddams & Co., Brisbane, 1914; Frank Cridland, *The Story of the Port Hacking, Cronulla and Sutherland Shire*, Angus and Robertson, Sydney, 1924; Reginald Harris, *Bivouac Tales*, Sydney, publication details unknown.

"A curious old gentleman": Charles Gogerly

On the northern shore of Port Hacking, only 1.5 kilometres across from where Gogerly supposedly saw the yahoo, is an inlet called Yowie Bay. In our previous book we dealt with that apparent "lexi-link" at some length but didn't come to any definite conclusions as to the derivation of the place name. (*The Yowie*, pp. 25-26)

In any case, since Gogerly's day many other people have reported encounters with hairy ape-men around Port Hacking, and one recent sighting occurred very close to the old Gogerly homestead. (Case 25, below, and *The Yowie*, pp. 26-27 and 284-5)

Case 25. 1860s – c. 1910. Near Sutherland, Heathcote and Waterfall, NSW

On 8 June 1907, in a letter to the *St. George Call*, an anonymous correspondent recalled how, one rainy night in the 1860s, he and some friends visited the abandoned village of Bottle Forest. As they took shelter inside the largest intact structure, Bottle Forest House, one man, identified only as "Pat" ventured outside to investigate a noise and returned to say that he'd encountered a "yahoo" that stood more than 12 feet tall. It had, he said, "starfish-like" feet.

Many local residents claimed to have seen the creature. They supposedly believed it could travel by land or water thanks to its webbed feet and suctioned under-soles, and that it could belch forth fire!

All very weird and wonderful – nice fodder for sceptics. On the positive side, however, the site of the old Bottle Forest village is only eight kilometres from where Mr Gogerly said he saw the giant yahoo in 1856.

Historian Patrick Kennedy records another incident from the 1860s:

"A Spaniard, who had a camp near Bottle Forest House … had a frightening experience one dark night, when one of his usually savage dogs ran to him in terror at the sight of a "hideous yahoo". [He] seized his gun and was frightened out of his wits when he encountered the "yahoo" moving among the trees.

Mr Kennedy notes that sightings were not confined to the immediate vicinity of Bottle Forest. From the late 1860s onwards, residents of the area between Sutherland (seven kilometres to the north) and Waterfall (five kilometres south) began hearing "strange and fearsome noises" at night from "The Thing", which was also known as the "wild man of the bush" and apparently resembled "a tall hairy creature that was neither man nor beast". At times its cry sounded "like someone screaming in pain". "The Thing" continued to make occasional appearances until about 1910.

St. George Call, 8 June 1907; Patrick Kennedy, "From Bottle Forest to Heathcote – Sutherland Shire's First Settlement" in *The Sutherland Shire Historical Society Bulletin* of July 1975.

Today, the Princes Highway, which links all the settlements between Waterfall and Sutherland, is effectively the western boundary of the 160-square kilometre Royal National Park. So it is interesting to note that contemporary residents of those settlements live on the very edge of the same wilderness that the colonial-era "Thing" supposedly roamed. Even more interesting is the fact that some locals claim to have seen similar creatures quite recently. (*The Yowie*, pp. 284-285)

Case 26. July 1861. Lake Grace, WA

Quoting *The Perth Enquirer* of the previous day, *The Courier*, Brisbane, of 26 Oct 1861, reported "the return of a volunteer exploring party, consisting of Messrs. C. And A. Dempster, B. Clarkson and C. Harper, with a native ... after penetrating nearly three hundred miles to the eastward of Northam."

While near Lake Grace, they were told by Aborigines that some years earlier "three white men, with horses", had arrived at another large lake further to the east, where they had either died of thirst or been killed by creatures called *Jimbars*. The natives described the creatures "as large, strong animals of the monkey tribe ... about the size of men ... very fierce – who will attack men when single, kill, and eat them. They also stated that [they] are not identical with the 'Ginka' or 'devil' of the natives, inasmuch as the 'devil' was never seen, while the others were both seen and felt by some of them."

The article concludes with the suggestion that a special expedition be sent to find the remains of the three unfortunate white men who, it was suggested, might have been remnants of Ludwig Leichhardt's 1848 expedition that disappeared while attempting an east-west crossing of the continent.

The Argus, Melbourne, 11 Oct 1861 and *The Courier*, Brisbane, 26 Oct 1861.

In 1888, while discussing this report, together with the 1851 Oakden and Hulkes story (Case 23) Ernest Favenc said that the Lake Grace Aborigines used the terms *jimbra* and *jingra* when referring to the monkey-men. He also made the point that "it must seem strange that the natives should in the *jimbra* have described an animal (the ape) they could not possibly have seen."

Ernest Favenc, *The History of Australian Exploration from 1788-1888,* Turner and Henderson, Sydney, 1888, p. 202.

Case 27. 1864 – Dec 1866. Murilla Mountain, 4 km south of Murrurundi, NSW

"STRANGE ANIMAL

For two years past a strange animal has occasionally been seen by people in the bush near the Murilla Mountain, and various have been the descriptions ... so that we have been doubtful ... but within the last two days two persons have seen the creature, which has caused much alarm to a whole camp of ... road-makers. It is described as being three feet six inches high, standing on its hind legs, the fore legs or arms could about touch the ground. It was covered with shaggy black hair ...

It made a horrible yelling when the parties rode in the direction of the rock it stood upon, showing a very fine set of teeth. It made a spring at its disturbers, who put spurs to their horses and fled. The blacks in this district are aware of ... these animals, and state that there were a great number of them some time ago. The place where [it] was seen is one of the wildest places that could be found ... at the back of the Murilla Mountain, or as it is generally called, the "Murlow". This creature evidently belongs to the ape type.

Kiama Independent, and Shoalhaven Advertiser, 3 Jan 1867. Also, *Sydney Morning Herald*, 11 Dec 1866; *Maitland Mercury and Hunter River General Advertiser*, 13 and 18 Dec 1866.

Case 28. c. 1860s. Clarence and Macleay Rivers region, NSW

"[The] blacks, many years ago, used to declare [that there] existed in the ranges between Rocky River (Timbarra) and the Clarence … [an animal] they called "Jerrawerra" … a biped, about the size of a small [woman], walking erect, and using its hands and arms as a human being. It was very rarely seen, and they [avoided] its haunts. They [say it lives] in caves, and [is always] ready to attack them … They also described them as slow in their gait, and covered with hair.

About 10 years ago I had a visit from Mr. _____, a squatter upon New England, a man who has been all his life in the bush, and who if I were to mention his name, it would serve as a guarantee anywhere for the authenticity of his assertions … he had seen and known all animals of common occurrence … and [not] likely to conjecture imaginary beings from an overheated brain. I asked him if he had heard about the "jerrawerras." He replied, "Oh; yes, often from old ____", naming a blackfellow I knew well. He then added, "I believe I saw one once myself."

[In the 1860s] He was travelling from ____ to his own place… night was closing in, but there was sufficient light to observe objects at a considerable distance. His track lay through … rough, unfrequented country, and … he had to descend a hill for about 300 or 400 yards, at the foot of which was a small creek, with tangled scrub … about half way down his horse pricked his ears, and exhibited unusual signs of interest in something ahead … and he saw what at the moment he took to be a blackfellow, walking among the bushes.

Thinking he was stalking some animal, he kept his eyes upon him as he approached. When within about 40 yards it quickly turned round, and, after gazing with astonishment at the man and horse, rushed into the scrub that lined the creek …

He says it was of low stature, not 4 feet high … it was not a human being, and yet resembled no Australian animal so much as human kind. He added, if it was not a "jerrawerra" he did not know what else to call it … The blacks at the station, when told of it, immediately pronounced it to be the "jerrawerra."

Clarence and Richmond Examiner and New England Advertiser, 31 Jul 1880.

Case 29. 1865 or 1866. Cordeaux River, NSW

In an article about an ape-like animal seen near Avondale in 1871 (Case 32) a brief reference is made to the sighting of a similar creature by a Mr B. Rixon at the Cordeaux River five or six years earlier.

The Illawarra Mercury, 14 Apr 1871; *The Empire* (Sydney), 2 May 1871.

Case 30. c. 1867. Braidwood, NSW

In his autobiography, William Derrincourt, an emancipated convict, told of his daughter's encounter with a hairy giant while lost in the bush near Braidwood. When writing our previous book, *The Yowie*, we possessed only a short excerpt from Derrincourt's book, and therefore couldn't accurately date the event.

Since then, however, we've accessed the entire book, and learned that not long before she became lost, Miss Derrincourt had had another memorable experience: an encounter with the notorious Clarke brothers.

Although the Clarkes and their henchmen committed many robberies and killed more policemen than any other gang of bushrangers, their reign of terror didn't last all that long: beginning in late 1865, it was all over by April 1867. It is likely, therefore, that the Miss Derrincourt's Hairy Man encounter occurred later that same year.

She told her father that on the third day of her ordeal she was confronted by, as she put it, "something in the shape of a very tall man, seemingly covered with a coat of hair and looking as frightened of me as I was of him. While he stood gazing at me without attempting to get nearer, I heard at a distance a peculiar cry, between a laugh and a bark, which my companion of the scrub answered in the same manner and, after seeming to consider for a few moments, he leisurely walked or shuffled off, greatly to my relief. I was afterwards told it was what the people here call a 'Yahoo' or some such name."

William Derrincourt, *Old Convict Days*. T. Fisher Unwin, 1899,
reprinted by Penguin, Ringwood, VIC, 1975.

Case 31. c. 1869. Avondale, NSW

Two children named Summers observed a monkey-like creature that was about the size of a 13 or 14-year-old boy. Two years later, census collector George Osborne encountered the same, or a similar creature in the same locality. (Case 32, below)

The Illawarra Mercury, 14 Apr 1871 and *The Empire* (Sydney), 17 Apr 1871.

Case 32. April 1871. Avondale, NSW

"A SUPPOSED GORILLA AT ILLAWARRA

We are indebted to the Illawarra Mercury for the following particulars supplied by Mr. George Osborne, of the Illawarra Hotel, Dapto, concerning a strange looking animal, which he saw last Monday, and which he believes was a gorilla. It is to be hoped successful means may be adopted to capture the animal (alive if possible), as it is quite evident it is one of the greatest natural curiosities yet found in the colony.

The following are Mr. Osborne's remarks concerning the animal: – On my way from Mr. Matthew Reen's, coming down a range about half a mile behind Mr. John Graham's residence, at Avondale, after sunset, my horse was startled at seeing an animal coming down a tree, which I thought at the moment to be an aboriginal, and when it got within about eight feet of the ground it lost its grip and fell. My feelings at the moment were anything but happy, but although my horse was restless I endeavoured to get a good glimpse of the animal by following it as it retreated until it disappeared into a gully.

It somewhat resembled the shape of a man, according to the following description: – Height, about five feet, slender proportioned, arms long, legs like a human being, only the feet being about eighteen inches long, and shaped like an iguana, with long toes, the muscles of the arms and chest being well developed, the back of the head straight with the neck and body, but the front or face projected forward with monkey features. Every particle of the body except the feet and face was covered with black hair, with a tan-coloured streak from the neck to the abdomen. While looking at me its eyes and mouth were in motion, after the fashion of a monkey. It walked quadruped fashion, but at every few paces it would turn around and look at me following it, supporting the body with the two legs and one arm, while the other arm was placed across the hip. I also noticed that it had no tail.

It appears that two children named Summers saw the same animal or one similar in the same locality, about two years ago, but they say it was then only the size of a boy about thirteen or fourteen years of age. Perhaps this is the same animal that Mr. B. Rixon saw at the Cordeaux River, about five or six years ago.

The Empire (Sydney) 17 Apr and 2 May 1871; *The Illawarra Mercury*, 28 Apr 1871.

Apart from the way that it remained on all fours, this smallish yowie has a lot in common with another, seen by H.J. M'Cooey near Ulladulla eleven years later. That creature, too, was grimacing like a monkey – "distorting its visage [and] blinking its eyes". (Case 61)

Case 33. 1871. Bulli Mountain, near Wollongong, NSW

"THE SUPPOSED GORILLA

A person who has resided on the Bulli Mountain for several years positively asserts that an animal similar to that seen by Mr. Osborne, but considerably larger, has been seen in the bush in that locality more than once, and by different persons, and that no dogs can be found to face it.

The Illawarra Mercury, 28 Apr 1871 and *The Empire* (Sydney), 2 May 1871.

Case 34. July 1871. Belgrave, near Kempsey, NSW

"ANOTHER GORILLA

It is said by persons [in] the neighbourhood of Belgrave, that a gorilla has made its appearance in that vicinity. A short time ago a camp of blacks were so scared by the appearance of the alleged monster, that they left their camp, and hastened ... to Warneton, and refused to return ... they said, "That fellow run on four legs, and stand up and run on two legs; him got plenty hair all over."

Two young men are also said to have been riding along ... between Belgrave and Warneton, when the supposed gorilla rushed through the bush near them, and so frightened [one horse] that it was with great difficulty that [the rider] could keep his seat, and prevent the horse from bolting.

A short time after … a person residing in the same neighbourhood, hearing his bull-dog barking and making desperate efforts to break his chain, evidently wishing to get at something he saw in the bush, let the dog loose. The dog, a very savage brute, immediately tore away in a furious manner … but in a short time was seen beating a speedy retreat, his courage evidently cooled. He took refuge in the house, and could not be persuaded to leave it.

A party of young men, it is said, formed a sort of expedition in pursuit of the alleged gorilla …

The Empire (Sydney), 20 Jul 1871; *Sydney Morning Herald*, 21 and 22 Jul 1871; *Queanbeyan Age*, 27 Jul 1871.

Many other Hairy Man reports have come from the Kempsey area since the 1870s. (e.g., Cases 148 and 190, and *The Yowie*, pp. 41-43)

Case 35. December 1871. The Jingera, part of the Gourock Range south of Captains Flat, NSW

"From the fastnesses of the Jingeras, adjacent to or in the district of Monaro, comes the startling intelligence that a "wild man" has been seen in that place.

A little girl, the granddaughter of Mr. Joseph Ward, senior, of Mittagong, asserts that she has met an old man, whose back is bent and body covered with a thick coat of hair, in height (to use the girl's words), about the same as her grandfather. The strange being in question had nails of a tremendous length on his hands and he seemed desirous of shunning the girl. The main points of the assertion are given with remarkable earnestness by Mr. Ward's grand-daughter; nothing can shake the simple outlines of her story.

Mr. Kelly, of the Jingeras, … says that he has himself seen the "wild man" … there is a tradition among the settlers of this place that the mysterious monster, the "yahoo", is a denizen of the mountain country … and that it is only observable in stormy weather, or on the approach of bad seasons.

The Monaro Mercury, 9 Dec 1871, *The Empire*, Sydney, 13 Dec 1871 and *The Queanbeyan Age*, 14 Dec 1871.

It is interesting to note that in 1904 Aborigines of the NSW south coast said that a mountain near Pambula, now known as Egan Peaks or The Jingera, was believed to be an abode of the Hairy Man/yahoo.

And as we have seen, a very similar term – *jimbra* or *jingra* – was used, in 1851, by South Australian Aborigines when referring to hairy ape-men in the vicinity of Lake Torrens. (Case 23)

Case 36. Mid to late 1800s. Tinderry Mountains, NSW

The term "jingera" cropped up again, when, in 1912, Colonel Granville Ryrie of Michelago spoke of a local belief in the Hairy Man.

> I first heard of the creature in my boyhood ... Some folks were dreadfully afraid of it. It was really the bogey of the district. It was supposed to be in the Tinderry mountains, and what is known as the "jingera" behind – the wild, rough country.
>
> No ape could live in [that] country. A carnivorous animal might get along, but there are no nuts, no yams, no fruits for a man-ape.

The Sun, 24 Nov 1912; *The Queanbeyan Age*, 3 Dec 1912.

Sir Granville Ryrie

Ryrie believed that the story was invented by cattle duffers to frighten people away. Despite his scepticism, the (omnivorous) creatures *were* credibly reported in the Jingera in the 19th century, and continued to be seen there into the twentieth. (e.g., Cases 35, 75 and 192)

Case 37. Mid to late 1800s. Braidwood district, NSW. Aboriginal lore

In 1912, in a letter to the *Sydney Sun*, Mr. A. B. Walton, of Granville, NSW, recalled hearing, many years earlier, Aborigines of the Braidwood district speaking of "big pfeller devil" that they called a *yahoo*. It was, they said, taller than a man, hair-covered and agile enough to climb trees. On one occasion, when disturbed by a party of blacks, it seized a woman by the throat and strangled her on the spot. Her companions fled in terror.

Mr. Walton was told that the fearsome creatures were only rarely seen.

> "I was only a boy at the time", he concluded, "and never saw the creature myself, but I have no doubt from what the blacks said that it is not a myth".

The Sun, 24 Nov 1912.

Case 38. December 1871. Crystal Brook, 20 km north of Proserpine, QLD

> We (*Port Denison Times*) have heard before of the existence, in the country around Mount Dryander and Crystalbrook, of an animal answering somewhat to the description of the gorilla, and of which the black fellows are said to be greatly afraid. We have hitherto paid but little attention to these rumours … Lately, however, we have been informed on the authority of a gentleman residing at Crystalbrook that two of these creatures have been seen. It would be a matter of considerable importance from a scientific point of view to know that this gentleman has not been mistaken, and if possible to capture one of these animals.

Darling Downs Gazette, 23 Dec 1871.

Case 39. Early 1870s. Ettrema Gorge, 30 km west of Nowra, NSW

A *Sydney Mail* correspondent wrote that settlers in the vicinity of Ettrema Gorge (now part of Moreton National Park) had long believed it to be the abode of Hairy Men. He quoted explorer John Chaffey as saying:

> "It may be 40 years back that a party of prospectors from the Shoalhaven ventured a few miles up the mouth of the Ettrema, and the first night they camped received a shock that hastened their movements.
>
> About midnight their dogs, with savage snarls and terrified looks, rushed into the tent and crouched beside their masters, then, making a bolt for it, cleared out; and the startled prospectors, looking out, saw what in the dim half moonlight they believed to be hairy men creeping around their tents. They were probably the large badgers or wombats which abound there.

Sydney Mail, 9 Oct 1912.

Case 40. March 1872. Lake Cowal, south-west of Forbes, NSW

This article is a strange amalgam of a bunyip (water monster) story and a yowie report:

> "THE BUNYIP AGAIN
>
> The "Bunyip" has again been seen twice within the last three months in the waters of Cowal Lake. In March last, by a party of surveyors, whose account can be relied upon, who were out in a boat and saw the animal about 160 yards off.
>
> They describe it as having a head something resembling a human being, or, in their own words, "like an old man blackfellow with long dark coloured hair." When seen it appeared to be going in a straight direction, rising out of the water so that they could see its shoulders, and then diving as in chase of fish, and rising again at intervals of about six or eight yards.
>
> They tried to get closer to it, but could not for the pace it was going, and consequently could give no description of it lower than the shoulders. They say the animal did not appear to be afraid of them.
>
> Another, a blackfellow, and a white man, who were out in a canoe, say they

saw it about a fortnight since. They agree in giving the same description of the head and hair as that given by the surveyors.

Sydney Morning Herald, 24 Aug 1872.

As we will see, Aborigines of the lower Murray believe that the Hairy Man, known to them as *mooluwonk*, is semi-aquatic (Case 145). Our files, in fact, contain several references to yowies entering water, swimming, or being encountered on islands. (e.g., Cases 11, 21 and 115)

Case 41. 1872. Braidwood district, NSW

"ANOTHER MYSTERY

For the past eight or nine days (says The *Braidwood Monitor*) an animal has made its appearance at the place known as the Giant's Cave, at the Pound Creek, which has caused nearly as much excitement as the much talked of but seldom seen bunyip. Some of the persons who have seen it pronounce it to be a monstrous wombat, whilst others say it is a native bear, while others again say it is a gorilla. However …none have been able to get within reach to have a cut at it with a stick or a shot from a gun.

It is to be seen generally of an evening sitting on one of the rocks in which this place abounds, and on the approach of any individual makes a chattering noise and immediately disappears amongst the rocks or into a hole, while those who have been in pursuit cannot find its secreting abode. Last evening there were many lads and others who were on the look-out for it with the hope of a capture – dead or alive.

Sydney Morning Herald, 16 Sept 1872; *Monaro Mercury and Cooma and Bombala Advertiser*, 21 Sept 1872.

Case 42. Mid 1870s. Pyramul, NSW

This item appeared in the *Lismore Northern Star Magazine* on 17 May 1878:

"A correspondent in a western district sends the following strange story to the "Freeman's Journal" :-

A few years ago young Tim Wring, a shepherd in Mr Price's employ … saw something unusual walking through the scrub but Tim could give no description, as he ran home for his life to be laughed at as a dreamer.

[Tim's encounter occurred about five miles from the place where the old shepherd saw the Hairy Man in c.1848.]

Later … in the [same] locality, Pat Wring, a younger brother, heard his kangaroo dogs bark from 10 a.m. to 4 p.m., down the inaccessible cliffs. He intended to go and help to kill what he supposed to be an old man wallaroo, as the dogs could kill any other kind of marsupial.

Pat's surprise may easily be imagined when he looked down on a hairy monster standing upright, a body apparently as round as a horse, arms as round as a man's thigh, three claws on each foot. It stood, to the best of his belief, about 4 feet high. The head resembled a pig's, but turned upwards, and it threw into the air the only dog that ventured within its reach. Pat could see the milk white hair under its armpits.

Pat … fearing the dog would be killed … threw about 14lbs weight of a stone, which struck the mark without doing any damage. The animal was at the foot of the rocks on which Pat stood, and … it sprang or strode in an upright position and then commenced to climb monkey-fashion.

Pat saw no more, as he thought it was time to run for his life; he never looked back. His heart beat so audibly that he fancied it was the quick stamping of the strange thing behind him. The dog died shortly after, but not a hair of the strange creature could be found, though the dog's hair and blood was plentiful on the rocks.

A *Freeman's Journal* article added further details:

It is now about 18 months since I first heard of Pat [Wring's] adventure. I thought … little about it … until I got the recital from his own lips. I fancy I am pretty sharp in detecting a falsehood … but I could see no reason to doubt the story. Moreover, the character of the whole family is above reproach. I have since seen the young man's sister, who tells me that when … Tim ran home … he was as white as a sheet …

She likewise reports another meeting with this strange thing. A settler's daughter having gone for the cows, an older sister, thinking she was long away, went out to assist her. On turning the corner of a bush fence, about a quarter of a mile from the hut, she suddenly stood face to face with the stranger.

No doubt both were frightened, as they stood watching each other, until the sister called out that she had all the cows, when the hairy creature turned about and walked leisurely away.

We always doubted the existence of this strange animal, but after conversing with some of the actors, and hearing the recital from neighbours who live beside them, we see no reason to discredit it any longer.

Freemans Journal, 13 Apr 1878; *Lismore Northern Star*, 17 May 1878.

Case 43. c. 1875. Creewah, near Nimmitabel, NSW

After reading about the Hairy Man seen by George Summerell near Creewah in October 1912 (Case 130), a former resident, James Allen, told of seeing a similar creature there, about 38 years earlier.

> It was very powerful-looking, covered with long, grey hair, and looked about sixty years old. It "was crawling around a steer", but made off into the forest when disturbed.

Zeehan and Dundas Herald (Tasmania), 8 Jan 1913.

Several yowie encounters have occurred near Creewah in the modern era. (e.g., Case 218 and *The Yowie*, pp. 66-67)

Case 44. 1875. Brindabella Mountains, NSW (now in the ACT)

> A party of horsemen, camped beside the Cotter River, discussed the possibility "of meeting a wild man, covered with dark shaggy hair, who, it is said, lives among these mountains."

The Brisbane Courier, 11 Dec 1875.

Case 45. April 1876. Crookwell, NSW

"THE GORILLA AGAIN

A correspondent, in writing to us from the Crookwell district, on the 23rd instant ... says:

The news [is] of what is supposed to be a veritable live gorilla, the monster having been seen within four miles of the village of Crookwell on several nights during last week.

Two or three families have been so much frightened by "the hairy man" that they have left their own houses and have gone to live together for safety. He went up to a window of one man's house, and looked into the room, showing his face, it is said, quite visibly, and scratching along the wall of the house with his nails ...

He is reported to be about six feet high and covered with shaggy hair, face and all. His tracks are said to have been followed for some distance, but [he] has not yet been found ... Last night a party armed to the teeth were going out to search for him ... It is said that shots have been fired at him without taking any effect on his body.

Illawarra Mercury, 28 Apr 1876.

Case 46. 1876. Walla Walla Scrub, 40 km west of Crookwell, NSW

"The Milbury Creek correspondent of the *Bathurst Free Press* says:- A resident of this place returned from the Fish River some forty miles from here, a few days ago, and told me that he had been informed by a respectable settler in that quarter that a party of sawyers, working in the Walla Walla scrub, came upon the dead body of an unearthly looking animal, human or inhuman they could not tell. It stood about 9 feet in height, with head, face and hands, similar to a man's; one of its feet resembled the hoof of a horse and the other was club-shaped; the body was covered with hair or bristles like a pig.

For many years past it had been believed by the settlers of that wild part of the country, that the Walla Walla scrub was inhabited by a monster commonly called "the hairy man of the wood," or what all the blacks stand so much in dread of – the Yahoo. Horses and cattle are said never to have been known to enter or remain in the scrub.

Australian Town and Country Journal, 4 Nov 1876.

Case 47. November 1876. Rocky Bridge Waterholes (at the present site of Abercrombie Reservoir), NSW

"WHAT WAS IT?

Who has not heard, from the earliest settlement of the colony, the blacks speaking of some unearthly animal or inhuman creature, that inhabited some part of the wildest, inaccessible, rugged and sequestered haunts of the rocky mountains and gorges in the colony, namely the Yahoo-Devil Devil or the Hairy Man of the Wood – which to this day they stand in awe and terror of.

I, for one, sympathise with the [Aboriginal belief] that such unearthly mongrels of monsters have and do now exist – though so rare ....

Fourteen days ago, and not more than ten miles from here, towards the head of the Lachlan River, on Coolamba station (Hammond's), in one of the most secluded and melancholy spots imaginable, imperceptibly a terror of awe creeps over every one that has to pass through this ... death-chasm of the river. While a lad of the name of Porter ... was shepherding a flock of his father's sheep, near to the dismal gorge ... an inhuman, unearthly-looking being [came] towards him from the high, rugged and precipitous rocks.

The dogs ... would not attack, became timid, and crouched around the lad's legs, who became horror-struck with fear; left the sheep to their fate, and ran, together with his collies, for home. On relating the inhuman sight [his story] was not credited by his father and others ...

On Saturday last, however, a fishing party of young men and women went to the Rocky Bridge waterholes [which] are famed, far and near, for quality and quantity of fish.

Two hours before sun-down, the young men and some of the women went to set their lines, leaving one of their young friends to boil the billy and prepare supper.

While engaged, the young woman was suddenly startled by observing a man [that] she naturally imagined … was one of their own party coming towards the fire, but on walking closer, discovered the appearance to be unsightly and inhuman, bearing in every way the shape of a man with a big red face, hands and legs covered over with long, shaggy hair, from fright became almost spellbound, screamed and screeched, but unable to run. The men, on hearing such unearthly cries, left their fishing lines and ran with all speed towards their comrade. On reaching the fire, the monster of alarm was only distant some fifty yards. On their appearing it stood for a minute or two and turned away and made for the rocks.

Two of the men armed themselves with a tomahawk and cudgel and followed this extraordinary phenomenon of nature for a short distance up the rocky and rugged mountain; when suddenly it turned round, and stood viewing [them] as they were approaching.

They also halted … about sixty yards from the object of terror, commanding a full view of his whole shape and make, resembling that of a big slovenly man. The head was covered with dark grissly hair, the face with shaggy darkish hair, the back and belly and down the legs covered with hair of a lighter colour. This devil-devil – or whatever it may be called – doubled round, and hurriedly made back towards the fire and women again. On seeing him coming, a fearful commotion amongst the females and a kind of supernatural terror amongst the men took place. In the meantime, before reaching the camp, it sidled away towards the inaccessible rocky mount.

The names of two of the men who … took part in the scene are Porter and Dunn, well-known settlers on the Abercrombie and Lachlan River. Mr Laner, another settler … has informed me … that the neighbours all round have organised a party to go in search of the human monster, and hunt him down, dead or alive.

It is well known to the old settlers for the last 30 odd years, that the blacks will never camp within miles of this death-like chasm of the Lachlan, though they come long distances to fish on adjoining waterholes …

Australian Town and Country Journal, 18 Nov, 1876; *Rockhampton Bulletin*, 1 Dec 1876.

On 27 December 2003, Dean Harrison of AYR received an email from a man who, while pig hunting with two friends seven days earlier, just three kilometres from the site of the 1876 incident, encountered "a medium size figure … a yowie/bigfoot/yettie, whatever they're called … the dogs were holding onto its shaggy fur … it shook [them] off … and ran into the forest … a few hunters from nearby towns have witnessed similar sights." (*The Yowie*, pp. 293-94).

Case 48. September 1877. Near Kiama, NSW

"THE GORILLA AGAIN

> … one day last week, a man living at Kangaroo Ground, whilst out in the bush saw what he believes … to be a gorilla. A party intend searching for this "wild man" of the forest … capturing him if possible, dead or alive. – Kiama Reporter.

Evening News (Sydney, NSW), 27 Aug 1877.

Case 49. October 1877. Sutton Forest, NSW

"AN EXTRAORDINARY ANIMAL

> Mr Prosser, manager at Messrs. Amos and Co.'s sawmills … has just informed us (Scrutineer) that a most peculiar animal has been seen by two men, Patrick Jones and Patrick Doyle, residents of Sutton Forest, in the bush between Cable's Siding and Jordan's Crossing.
>
> Mr Prosser himself has seen the footprints; they are 3 feet apart, and the impression made by the feet is similar to that of an elephant. The animal is described as being 7 feet high, with a face like a man, and long shaggy hair, and makes a tremendous noise. Fourteen of the men from the mill, fully armed, intend starting on Saturday next to endeavour to capture this "wild man of the woods." Mr Prosser assures us there is no exaggeration about this affair, and every one at the mill believes in the existence of this strange creature.

Sydney Morning Herald, 12 Oct 1877.

Case 50. Late 1870s. Near the Laura River, north QLD

There's not much doubt that the following tale, which was published in *The Clarence and Richmond Examiner* on 21 March 1893, after first appearing in *The Australian*, is completely bogus.

Titled "The Birrinji-Wallah" and written by someone calling himself "Majuba", it was apparently considered, by the editors of *The Australian*, who allowed it to run for eleven mind-numbing pages, to be an absolute "ripping yarn." We, however, would like to just rip it up.

But as it contains what purports to be north Queensland Aboriginal yowie lore, and features the supposed abduction of a prospector by one of the hairy beasts, we feel obliged to include it (heavily edited) in our collection.

The episode supposedly occurred during the Palmer River gold rush of 1872-79, during which hundreds of north Queensland Aborigines and scores of prospectors were shot or speared to death in bloody skirmishes.

"Majuba" begins by saying the local Aborigines "have always been in great dread of some monster they call 'Birrinji-Wallah' … a wild man, covered with coarse hair, and possessing the strength of a gorilla".

He claimed that as he and his mates were prospecting one day, a gaunt, half-crazed man, dressed in rags, ran towards them from the jungle.

When he calmed down, the man, Peter Nash, said that during an ambush at the Laura River he'd been knocked unconscious and carried off by Aborigines, who he feared were cannibals. On reaching their camp however, he had the presence of mind to pull out his "pocket looking glass" and dazzle a warrior who was about to club him to death. His captors, fearing that he possessed supernatural powers, then left him to recuperate in peace.

The peace didn't last long: one morning, when the warriors were away, all the women and children suddenly ran from the village screaming. Turning around, Nash saw something that "caused a cold shiver to pass through him".

"Stalking through the middle of the camp came a hairy monster, with a stealthy and noiseless tread. Erect on its hind legs, it stood fully 6ft. high, and exactly resembled pictures he had seen of the gorilla. [It] glared round … and sent forth a roar which seemed to shake the ground [and] came towards him, snarling and showing his fangs."

Petrified with terror, the poor bloke was then abducted again – lifted like a baby and hauled off into the scrub. It was all too much – "he went off into a swoon" and only awoke as the creature, still carrying him, descended a sheer cliff by way of overhanging branches.

Keen students of Canadian sasquatch lore might find the next bit interesting: the creature (like the bigfoot that supposedly abducted Albert Ostman in 1924) had taken him to a small valley, entirely hemmed-in by cliffs, where it treated him "as a boy would keep a rabbit … as something to look at occasionally", but generally ignored him. There was nothing to eat other than wild plums.

After escaping by way of a narrow fissure in a cliff and wandering for days, he stumbled into "Majuba's" camp and blurted out his story.

A ripping yarn, to be sure. But wait – there's more:

Nash revealed that, before escaping, he'd discovered that the hidden valley was shot through with reefs of gold. Naturally, "Majuba" and party resolved that as soon as he recovered, they'd accompany their visitor back his "El Dorado", kill the monster and grab the gold.

But (wouldn't you know it?) their plans were thwarted – two nights later, "a most extraordinary" bellowing was heard from the scrub, and Nash ran off into the night.

"Majuba" and co. searched for a couple of days, gave him up for dead, and returned to camp, only to find it had been raided by Aborigines, who'd taken most of their food and, just for luck, beheaded their mate Murphy, who'd been standing guard.

Burying poor Murph, and suddenly wracked by deadly fever, they tottered off and "with difficulty some of us managed to reach the coast."

Case 51. Late 1870s. Between Tooraweenah and Mendooran, NSW

In 2015 we were contacted by Grant Baillie, who had, while undertaking genealogical research, come across an interesting entry in his great grandfather's diary.

It detailed something that occurred at a place called Log Hole, on a stock route near what is now Goonoo State Forest: a dramatic encounter with what he called a "clubfoot" or "gorilla".

We never sighted the original diary entry, but the text, as conveyed to us by Grant, is as follows:

> "They left Molong and came up and joined Tom Hogben. He was camped at Log Hole, a few miles down from Tooraweenah. My father and two of the Hogben girls went kangaroo hunting. They were coming home just on dark. They saw something over a bit. Father had a big staghound dog that would tackle anything. He said, "get him".
>
> The dog rushed over. The thing made a grab at the dog [which ran off with its tail between its legs]. The girls let out screams and ran. He threw the kangaroo skins and [ran] after them. Tom Hogben was a very good tracker. He went down next morning. He said it was clubfoot or gorilla. Jim Page was supposed to have shot one in that area later.

Case 52. 16 February 1880. Yorke Peninsula, SA

Were it not for its supposed covering of feathers – and, perhaps, the oddly-shaped footprints, this creature would conform pretty well with those described by modern-era witnesses.

> RATHER SENSATIONAL.
>
> Our Minlaton correspondent, writing on 17th February, says: A singular-looking object was seen yesterday, at twenty minutes to 7 p.m., by Mr. M. O'Dea, a well-known resident [when] driving from Stansbury, and about one and a half miles from Minlaton, the horse gave a sudden start, and on looking up he saw a strange-looking object on the road, about two and a half chains [50 metres] in front of him.
>
> [It] was formed like a man, stood from 7 feet to 7 and a half feet in height, and was covered with feathers of a very light colour. It went along the centre of the road for about 200 yards at a great pace, taking enormous strides.
>
> Mr. O'Dea put his horse into a gallop, but it still increased its distance until coming to a bush fence, when it put out one of its arms, which was like that of a human being, only very much longer, and made a loud noise, which he could only describe as being between the bellow of a cow and the peculiar noise made by an emu when calling for its young. It easily cleared the fence, and was soon lost to view in the thick scrub … the outline of its face was at all appearance like of a monkey's.

Advocate (Melbourne), 1 May 1880. Also *The South Australian Advertiser*, 20 Feb 1880.

On 4 May the *Advertiser* ran a follow-up story by its Minlaton correspondent:

> Hearing that the tracks of the animal had been seen on the farm of Messrs. Trout & Smith, about six miles from where Mr. O'Dea saw it … Mr. Smith supplied me with the following particulars:-
>
> About a week ago a great noise was heard among the farm horses in the middle of the night … next morning … it was found that … Two had broken their ropes and another had strained its rope so tightly in its efforts to break loose that it was with difficulty untied.
>
> … tracks of some large animal were … followed from a slip-panel on one side of the house to a water cask, from which it had to all appearances drank. From there [they] were followed for nearly a quarter of a mile to the scrub. Mr. Smith

describes the tracks as being very plain, in appearance like a man's hand, only very much larger … about twice the size of an ordinary hand.

The length of the stride … was … four feet six inches [1.37 m]. The weight … must be great … the toes … seemed to have sunk in to the depth of from an inch to an inch and a half. Mr. Smith, who is an old bushman and familiar with all the known animals of South Australia, has never seen any tracks … l resembling these.

It seems evident that something is living in the scrub which is entirely unlike anything we know of. The early settlers here say that the aboriginals used to maintain that there was "one big fellow like a man along a scrub." There is some talk of a party being organised to make a search, but as there is over seventy miles of dense scrub to explore it would be no easy or pleasant task.

The Maitland Mercury and Hunter River General Advertiser 28 Feb and 4 May 1880 (quoting *The South Australian* Advertiser, 20 Feb 1880).

Case 53. May 1880. Near Nymboida, north-east NSW

"From the far north comes a story of a wild man of the woods of an indescribable appearance … Some such creature is reported to have been seen not far from Nymboida, in this district, and about which there are some astounding tales, many of them of the Munchausen character.

… might not some of our sporting men make up a party to scour the country in that direction. If they did not succeed in discovering the "hairy man", they might find something much more valuable. The locality … will be none the worse for careful exploration for more reasons than one.

Clarence and Richmond Examiner and New England Advertiser (Grafton), 22 May 1880.

Case 54. July 1880. Serpentine Creek, about 10 km south of Ebor, NSW

Apart from having, supposedly, a bizarre, snake-like head, the creature described in the following article seems generally yowie-like.

GRAFTON

Tuesday.

Constable Morgan, of Blicks River, reports to sub-inspector Creagh that several well-known residents have seen an extraordinary creature in the bush on the Serpentine River, near the Grafton and Armidale Road. The creature is described as about three feet six inches in height, with a head like a diamond snake, hands like a kangaroo, and a body like a man's, but dark and hairy. One man saw the creature on six different occasions, and on one occasion he was chased by it.

Sydney Morning Herald, 7 Jul 1880.

On 24 July the *Clarence and Richmond Examiner and New England Advertiser* provided a follow-up:

Senior Constable Morgan … was in town this week, and … he informs us that [the creature] has been seen by two parties, one of whom, a boy of twelve years, has come across it repeatedly. On one occasion he was so close as to be able to determine that it was of the male sex; he threw a stick at it, when it gave chase to him, but its progression is so slow that the boy, who was on foot, could run faster …

He describes it as between 3 and 4 feet high, [walking] erect like a human being, having very thick thighs, and covered with dark brown hair. It is always seen about the one locality, but never in the winter time … a party will be organised to [capture it], and Morgan is very sanguine of success ….

Case 55. May 1881. The Jingera (part of the Gourock Range south of Captains Flat) NSW

The Cooma Express relates that the Jingera hairy man has again turned up. It was seen on Saturday last by Mr. Peter Thurbon and one or two others. This is its first appearance for some considerable time past. The animal, if such it be, has the appearance of a huge monkey or baboon, and is somewhat larger than a man.

The Goulburn Herald, 24 May 1881.

Case 56. August 1881. Near Inverell, NSW

> Something, supposed to be a gorilla, was seen last Sunday in the neighbourhood of the Big River by two men while out fishing, who returned and reported it to the police. A search party went out yesterday, but have not yet returned.

Armidale Express and New England General Advertiser, 2 Sept 1881.

On 9 September, the same paper stated:

> One of the party, who came back on Thursday reports that up to the time of his leaving, nothing had been discovered except some tracks, about the size of a small man's foot but different in shape.

Case 57. October 1881. Mount Macdonald, NSW

> On the eastern side of this goldfield lies a large tract of mountainous country formed of granite. In this little known country a wild man is known to have existed … He has been seen frequently, and is said to be covered with hair. Although pursued on many occasions he has always succeeded in escaping.

The Argus (Melbourne) 5 Oct 1881.

The village of Mount Macdonald is now very close to the northern shore of Abercrombie Reservoir, just downstream from Rocky Bridge waterholes – site of the dramatic Hairy Man event of 1876. (Case 49)

Case 58. April 1882. Near Augusta, WA

> A rather remarkable report has come in from Augusta, to the effect that a gorilla has been shot in that neighbourhood by Mr. Elliot Brockman, and that another one had been seen … several persons coming from Augusta vouch for the correctness of the fact. Where could these two animals (accepting the story as true)

have come from? They must be survivors from some wreck on the coast not reported.

The West Australian, 14 Apr 1882.

Twelve days later another West Australian newspaper carried a sceptical follow-up:

"Elliott Brockman has shot a gorilla … and another … has been seen. Such is the rumour, which adds, that the height of the slain individual is no less than 8ft. 6in. The Busselton correspondent of the West Australian says they must be the survivors from some wreck on the coast – not reported. He is, very probably, correct – so many vessels come cruising along our shores, laden with eight-foot gorillas, and other rarities of the animal world.

Victorian Express (Geraldton, WA), 26 Apr 1882.

Case 59. June 1882. Near Keera, 45 km south-west of Inverell, NSW

"The *Inverell Times* gives the following … from two young townsmen, Messrs. James Skinner and F.T. Booth [who] went on a fishing excursion … at the Big River [now the Gwydir?] at the bottom of the bite, about a mile and a half from Mr. Browne's selection on the Kera [Keera?] run.

As they were fishing – at about half-past 12 o'clock – they were surprised to see, floating against the stream, two corks; while another cork was observed to splash in the water, about fourteen yards off. Scarcely a minute elapsed before a second splash occurred, bigger than the first.

Under the impression that it was a fish … Skinner took his line to the spot; he had been there about five minutes, when Booth saw a stone skip the water, about ten yards from where he was.

After skipping the water, twice, the stone sank near the opposite bank. A minute afterwards Skinner heard a stone whizzing, which must have been a pound weight, and which passed close to his head. He very naturally "ducked" … Skinner very narrowly escaped another stone.

Skinner … saw something [about 20 yards away] which made him cry out, "Oh my God, we are done; there is a gorilla." Booth shouted, "Don't run away;

let's see what it is", and jumping on a neighbouring rock he saw what he believed to be a gorilla. The unknown rapidly disappeared, however, among some ti-trees, rocks, and long grass.

[Skinner] describes the body as being apparently covered with hair; and with head and ears resembling a monkey's. It stood upon two feet, and had arms and legs. The eyes, he says, were very large – as big as a two-shilling piece. [Booth said that] the back was completely covered with brown hair, fully five inches long.

The travellers … turned … to get their horses. As they were leaving … two stones were thrown at them; these fell … about twenty feet distant …

Our adventurers then rode to Maidswater – about two miles off, for further fishing. They were there about an hour when Booth heard a low cry in the scrub about one hundred yards off.

"Did you hear that?" he said to Skinner … "I think he is here again … You had better come home". Skinner complied, but … had forgotten his pipe [and] returned to the river … when another stone lodged on the ground about five yards from him. This stone (a parting shot) Skinner believes must have weighed five pounds.

Nothing further occurred out of the ordinary … and the young men returned immediately to Inverell, where they reported the foregoing strange occurrence to the police.

We believe that a search party is to be organised … and great curiosity will be naturally felt to see the one described in the foregoing narrative.

Illawarra Mercury (Wollongong), 23 Jun 1882.

Case 60. 14 October 1882. Treachery Head, near Bulahdelah, NSW

"WHAT WAS IT?

A correspondent writes as follows to the editor of the Echo:

"Last Saturday an animal was seen from the beach, by the aid of a telescope, in a sitting posture on Treachery Headland, by five persons at Seal Rocks. Attention was first drawn to the subject by the hearing of a crying noise similar to that of a child. It is said by one to be hairy. This is now the third creature seen in these parts resembling the hairy man. One was seen at Boolambayte; another at the Upper Myall, off Port Stephens."

The Clarence and Richmond Examiner and New England Advertiser, 21 Oct 1882; *The Mercury*, Hobart, 17 Oct 1882.

Case 61. December 1882. Between Batemans Bay and Ulladulla, NSW

Letter from H. J. M'Cooey to "The Naturalist", *The Australian Town and Country Journal*, 9 Dec 1882:

"AUSTRALIAN APES

Reports occasionally reach us … of strange animals of the monkey tribe being seen in different parts of the colony …

A few days ago I saw one of these strange animals in an unfrequented locality … between Bateman's Bay and Ulladulla. My attention was attracted … by the cries of a number of small birds which were pursuing and darting at it. When I first beheld the animal, it was standing on its hind legs, partly upright, looking up at the birds … blinking its eyes, distorting its visage and making a low, chattering kind of noise. Being above the animal on a slight elevation and distant … less than a chain [about 20 metres], I had ample opportunity of noting its size and general appearance.

… if it were standing perfectly upright it would be nearly 5ft high [1.52 m]. It was tailless and covered with very long black hair, which was of a dirty red or snuff-colour about the throat and breast. Its eyes, which were small and restless, were partly hidden by matted hair that covered its head. The length of the fore-legs or arms seemed to be strikingly out of proportion with the rest of the body, but in all other respects its build seemed to be fairly proportional.

It would weigh about 8st. [50 kg]. On the whole it was a most uncouth and repulsive looking creature, evidently possessed of prodigious strength, and one which I should not care to come to close quarters with. Having sufficiently satisfied my curiosity, I threw a stone at the animal, whereupon it immediately rushed off, followed by the birds, and disappeared in a ravine which was close at hand.

I do not claim to be the first who has seen this animal, for I can put my finger on half a dozen men at Bateman's Bay who have seen the same, or at any rate an animal of similar description ... I may mention that a search party was organised at Bateman's Bay some months ago to surround the locality the supposed ape frequents and shoot or capture it ... I may further state that the skeleton of an ape, 4ft in length, may be seen at any time in a cave 14 miles from Bateman's Bay, in the direction of Ulladulla.

Yours truly, H.J. M'COOEY

H. J. M'Cooey

Henry James M'Cooey (1852-1902) was a journalist and also a collector for the Sydney Museum. Although his statement about the readily accessible ape skeleton is very dubious, the site of his alleged encounter accords well with local Aboriginal lore.

In fact, the area between "The Bay" and Ulladulla, is still something of a yowie "hot spot." (e.g., Cases 183 and 193)

Case 62. Early 1883. Between Mount Keira and Mount Kembla, NSW

> "It is stated by several persons residing in the locality that a gorilla has again been seen in the mountain ranges in the vicinity of Mount Keira, or rather, between Mount Keira and the Mount Kembla coal tunnel. Different residents there aver that they have caught sight of the strange animal on separate occasions and in various places. They describe it as resembling a man, but covered with long hair, and having long sharp claws.

The Illawarra Mercury, 19 Jan 1883.

Hair-raising encounters with gorilla-like animals still occur on and around Mt. Keira and Mt. Kembla. (e.g., Case 210 and *The Yowie*, pp. 183, 211, 253 and 289)

Case 63. 1883. Between Keera and Cobbedah, north of Barraba, NSW

William Telfer

New England pioneer William Telfer (1841-1923) was a self-educated man who wasn't overly concerned with the niceties of spelling and punctuation. This is an extract from his memoirs:

> "i had an Experience of this gorilla or hairy man in the year 1883 i was making a short cut across the bush from Keera to Cobedah via top bingera ... made a camp on a high bank of the creek lit a fire and made myself comfortable my dog laying down at the fire alongside me.
>
> i sat smoking my pipe the moon rose about an hour after when you could discern objects two hundred yards away ... i heard a curious noise coming up the creek opposite the camp over the creek i went to see what it was.
>
> about one hundred yards away he seemed the same as a man only larger ... something like the Gorilla in the Sydney museum of a darkish colour and made a roaring noise going away towards top Bingara the noise getting fainter as he went along in the distance.
>
> i started at daylight Getting to Bells mountain at about 9 oclock Mr Bridger lived there stopped and had breakfast ... they said several people had seen the gorilla about there he was often seen in the mountains towards the Gwydir and about mount Lyndsay.
>
> i was thinking how Easey this animal could Elude pursuit travelling by night camping in Rocks or Caves in the daytime ... they are faster than the aboriginal by his own account. some people think they are only a myth but how is it they were seen by so many people in the old times fifty years ago.

William Telfer, *The Early History of the Northern Districts of New South Wales*, c.1898, University of New England Archives A147/V213, pp. 32-34.

Mr. Telfer's encounter may have occurred less than ten kilometres from the spot where Messers. Skinner and Booth were harassed by the rock-throwing "gorilla" a year earlier. (Case 59)

Case 64. September – October 1883. Between Bathurst and Orange, NSW

> "A strange animal like an ape has again been frightening the residents on the road from Orange to Bathurst, at a place called The Rocks. Several ineffectual attempts have been made to secure it.

Evening News (Sydney), 2 Oct 1883; *Australian Town and Country Journal*, 6 Oct 1883.

> A party of horsemen started this morning to hunt for a strange animal resembling a baboon, which for some time past has been a great source of alarm in the district. It was again seen the other day near the Bathurst road.

Evening News (Sydney), 26 Oct 1883.

Case 65. Late 1883. "Barringdun", near Nanango, NSW

This is one of the very few reports in our files that refer unambiguously to human beings-gone-feral. (See also Case 85)

> "We learn from a Nanango correspondent that a naked man was captured on Barringdun run last week, after a determined and exciting chase. He had been seen from a distance in various parts of the Burnett district, at intervals during the past six months.
>
> His captors conveyed him to the Nanango lockup, whereupon he stated that he had been living in the bush, destitute of clothes or ordinary food, for the last four months, and relied for sustenation [sic] on what he could pick up in the bush. He assigned no reason for leading such a reclusive life. The man proves to be one O'Neil, a stockman, under suspicion of horse-stealing.

Brisbane Courier, 12 Jan 1884.

Case 66. c. 1884-85. Bumberry, 20 km east of Parkes, NSW

After reading of a "gorilla" sighting near Braidwood a sceptical correspondent wrote to the *Australian Town and Country Journal* to say that similar events had been reported in the Bumberry region prior to 1885:

> "On several occasions the "hairy man" was seen in these parts by various individuals, and nearly always when they were engaged (as the party near Braidwood) in boiling their billys for supper ... People have generally described him as a hairy creature. But no one seems to have remained long enough ... to ascertain his exact height.
>
> Three thousand years ago a celebrated warrior king said in his haste, "All men are liars." In these days it is a self-evident fact that there is quite a host of story-tellers in New South Wales.

Australian Town and Country Journal, 1 Oct 1887.

Case 67. July 1884. Near Cudgegong, NSW

> "A monster, said to resemble in size and shape a huge gorilla, makes frequent appearances about Cudgegong. Some men camped in tents near the Rhobardah gold mine were greatly frightened by the entrance of the "wild man of the woods" into one of their tents, but they cannot describe it, as they all left as speedily as wind and limbs would allow. Something similar was said to have been seen near Blayney about six months ago.

The Maitland and Hunter River General Advertiser, 10 and 12 Jul 1884.

Case 68. April 1885. The Gwydir River, south of Inverell, NSW

In Case 59 we related the story of two fishermen who encountered a rock-throwing yowie on the banks of the Gwydir (then known as Big River) in June 1882.

Three years later, the critter, or a similar one, returned:

The "Hairy Man"

HE TURNS UP AGAIN. EXCITEMENT IN INVERELL.

A young man named McRae informs us that on Good Friday, when fishing at the Big River, he noticed an object apparently swimming ... towards him. He threw a stone, and the figure turned away. He then took his gun and fired at it. Upon this the mysterious stranger sprang about 6ft into the air, and rapidly swimming to the opposite shore, disappeared in the bush. McRae describes it as black and hairy, and of a shape hitherto unknown to him! Query: can there be a gorilla in these parts?

Evening News (Sydney), 2 May 1885 (quoting the *Inverell Argus* of 25 April); *Maitland Mercury and Hunter River General Advertiser*, 23 Apr 1885 (quoting the *Inverell Herald* of 18 Apr); *Australian Town and Country Journal*, 2 May 1885.

Case 69. c. March 1885. Pearce's Creek, NSW (now ACT)

After journalist John Gale published an exaggerated version of the incident, George Webb of Uriarra wrote to the *Queanbeyan Age* on 7 Aug 1903 to describe exactly what happened when he and his brothers, along with John and Alexander McDonald, encountered a strange animal about eighteen years earlier:

WAS IT THE HAIRY MAN "YAHOO," OR THE WILD BLACKFELLOW OF THE ABORIGINALS?

IF IT WAS NEITHER, WHAT WAS IT?

Sir, – I send you an account of what took place when some strange animal came very close to where my brothers and I were camped many years ago.

We were out in Pearce's Creek in search of cattle ... decided to [camp for the night]. The weather was very hot and dry ... there was no moon, none of us had a match. We had supper as usual, and lay down.

Sometime during the night ... I heard a noise similar to what an entire horse makes. I heard it again and awoke the others. We heard it some four or five times, and the noise ceased, but we could hear it walking along on the opposite side of the range, and when in a line with our camp, we could hear it coming down in our direction. As it came along we could hear its heavy breathing. About this time the dogs became terrified and crouched against us for protection.

... it had to come [around a fallen tree] to get to where we were. Not many seconds passed before Joseph sang out, "here the thing is," and fired a small pistol ... at it. Neither William nor myself ... got a sight of it. Joe says it was like a blackfellow with a blanket on him.

We did not hear it going away. We then tried to set our dogs after it, ... but could not get them to move.

Yours etc., Geo. Graham Webb. Woodstock, Uriarra.

Although only Joseph got a (not very clear) look at it, the nocturnal prowler's upright stance, its heavy tread and heavy breathing – and the extreme fear it triggered in the dogs – strongly suggest it really was a yowie.

The youngest member of the party, Alexander McDonald, seemed to believe so. In later life, he gave his son Eric a sketch that was drawn "from memory" by one of the older men.

The Pearce's Creek yowie, as sketched by "an eyewitness".

William Webb

Joseph Webb

Although it is obvious the artist gave free rein to his imagination – the spurs, the horns, the cheerfully grinning visage – while depicting the creature, Alexander kept the sketch to the end of his long life, and insisted the story was true.

John Gale, *An Alpine Excursion,* Queanbeyan, 1903, pp. 85-89; *Queanbeyan Age*, 7 Aug 1903; handwritten account given by Eric McDonald to Lyall Gillespie, 10 Mar 1979.

Case 70. 1885. Parkers Gap, southeast of Captains Flat, NSW

According to historian Netta Ellis, "*The Town and Country Journal* of 1885 reported that a hairy man or orang-outang had been seen at Parker's Gap in the Gourock Range."

Netta Ellis, *Braidwood, Dear Braidwood*, p. 161.

Case 71. August 1885. Near Tarago, 14 km east of Lake George, NSW

"THE HAIRY MAN AGAIN

> No little commotion has been caused … during the last few days by the rumour that the hairy man has been seen in the high range at the rear of Mr. William Wyatt's farm. Wallaby hunters declared with solemn, scared faces that they had had a glimpse of him … one of our local residents, a giant among men, states that there is no doubt about the identity of the mystery; and on Tuesday night last a settler and a lad went in that direction resolved to do or die.
>
> It appears that his lordship was seen in the moonlight … and appeared about eight feet high, covered with hair, and when fired at roared as loud as distant thunder, the shot taking no effect beyond making his eyes, plainly visible, sparkle and glare, while something like fire shot out of his nostrils. The settler states he walked backward until able to get a fair run from the monster, while the youngster reached home with a scared, white face.
>
> This tale is told in front of the bars to eager, and sometimes incredulous, listeners; but tomorrow a posse of bold hunters will invade the solemn, dark looking ranges, carrying their lives in their hands, a trusty rifle, and a bottle, perhaps … Time will prove whether it is an old man kangaroo, or cross between a kangaroo and a bull.

Goulburn Evening Penny Post, 22 and 29 Aug 1885; *Australian Town and Country Journal*, 29 Aug 1885.

The jokey mention of gunfire causing the creature's eyes to "sparkle and glare" is interesting in view of the many references to brightly glowing eyes in modern-era reports.

It is worth noting that Bungendore resident Billy Southwell experienced a dramatic, face-to-face encounter with a smaller yowie at "Currandooley," just 18 kilometres south-west of Tarago, in 1976. (*The Yowie*, pp. 234-35)

Case 72. 1886. Falconbridge, NSW

> "… about three years ago … the peaceful little village of Lindon was shaken to its centre by the report that a monstrous and mysterious apparition had appeared to a lady, the wife of the caretaker of Sir Henry Parkes's property at Falconbridge.

It seems that she was … gathering a few sticks when a commotion amongst the fowls attracted her attention, and on looking up, before her stood a Thing about seven feet high. The black hair growing on its head trailed weirdly to the ground, and its eye-balls were surrounded with a yellow rim. It was – the hairy man!

Now one would suppose that the reflected glory from so great a man as Sir Henry would have lent the lady courage to face the monster … but no, she just dropped her sticks and "skinned out of that," giving utterance to piercing screams.

Her husband … sallied forth with his gun to put daylight through him; but he … had disappeared, taking several of the fowls with him … But, and this was imparted to us with great solemnity, he left a track three inches deep behind him!

Illustrated Sydney News, 3 Oct 1889.

Case 73. January 1886. Koetong, 30 km east of Tallangatta, VIC

"A report has been received at Albury from the Victorian Upper Murray that a wild man, perfectly nude, and covered with long hair, has been seen in the ranges near Koetong. A search party, in company with mounted troopers, is searching the locality.

The Brisbane Courier, 25 Jan 1886; *Australian Town and Country Journal*, 30 Jan 1886.

Case 74. July 1886. Branxton, NSW

"A friend writes from Branston [now Branxton] in the Upper Hunter district, that – "One of the selectors reports that while cutting timber he saw what he took to be a great monkey walking upright like a man. The thing had enormous arms, great width of chest and shoulders, and a white stripe down the middle of the face. Since then the creature has been seen in the Pokolbin Hills, some 20 miles from where first seen. All the observers say that its appearance was horrible, and that it gave utterance to a most appalling roar or articulation.

> Besides, tales are current in that district of the blackfellow's dread of the "big one wild fellow" so that sufficient data are presented to justify a search expedition.

Ovens and Murray Advertiser (Beechworth), 24 Jul 1886.

The Pokolbin Hills are in the vicinity of Broke – where a remarkable series of yowie sightings occurred in 2015-2017. (Case 254)

Case 75. August 1886. The Jingera, east of Bredbo, NSW

"THE JINGERA YAHOO

> Whilst a young man named Flynn was looking after stock at the back of the Bredbo station one afternoon last week, he was surprised to observe a hairy human form, about seven feet in height, walking in the bush. The wild man walked with an unsteady, swinging, and fast step, his arms being bent forward and nearly reaching the ground, whilst the colour was described as "bay," between a red and chestnut.
>
> Flynn … rode as fast as he could to the homestead of Mr. Crimmings … . Since then, Mr. Crimmings himself has interviewed the monster, and his account tallies exactly with that given by Mr. Flynn. But Mr. Crimmings heard the animal make a cry that sounded very like "Yahoo."
>
> We hear that Mr. Joseph Hart, of Jingera, also saw the "Yahoo" as he was returning home one afternoon. The strange being is, no doubt, the "Wild man" that has been so often talked of about Jingera for so many years past. It is the intention of Bredbo and Jingera residents to scour the bush in a strong body and capture the monster alive or dead. For this purpose they will meet at Mr. Kelly's hotel Little Plain on Monday next.

Queanbeyan Age, 24 Aug 1886.

The Jingera certainly was – and continues to be – something of a yowie hotspot. (e.g., Cases 35, 55 and 192)

Case 76. September 1887. Gilberts Creek, 6 km south-west of Braidwood, NSW

"ANOTHER YAHOO

> About a fortnight since, one John Mahony, in the employ of Mr. Thomas Lee, who has a contract for erecting a bridge at Gilbert's Creek, on the road to Cooma, and about a couple of miles out of town, was … cooking his and his mate's supper just after dusk, when he saw a hairy individual, 7 feet high at least, marching down without the least concern for anybody, and striding across 5 feet drains and 5 feet high fallen trees without the slightest trouble, and proceeding on his way wholly oblivious of anything around him.
>
> John Mahoney cleared without asking any further questions of the strange intruder. It was a full moonlight night, and the figure was not more than twenty yards away from him. Numerous stories of the same kind are current … and we could mention names to show the credibility of our information … The appearance is described as that of a gorilla, about 7 feet high, all hairy from head to toe, and of a light colour.
>
> Other persons have seen the creature, whatever he be, in various parts of the district, viz., at Monga, Parker's Gap, and the Sassafras, in every instance to their utter terror, most of them being carriers, who avow that they will never pass over the same part of the road again unless in company with someone else.

Goulburn Evening Penny Post (quoting the *Braidwood Dispatch*), 17 Sept 1887; *Australian Town and Country Journal*, 17 Sept 1887; *Queanbeyan Age*, 1 Oct 1887.

As noted earlier (Case 30) Miss Derrincourt encountered a Hairy Man near Braidwood in c.1867. Sightings have also occurred, in the modern era, on Clyde Mountain, about 20 kilometres south-east of Braidwood. (Case 206, and *The Yowie*, pp. 57-58)

Case 77. June 1888. Between Glen Innes & Grafton, NSW

"THE "YAH-HOO"

> The whole countryside between Glen Innes and Grafton has … been kept in

a tremor of apprehension by oft-repeated awful roars ... by some mysterious monster which, for want of a better name, is called the "Yah-Hoo".

One individual recently came into town ... and reported that he had actually seen the "Yah-Hoo." Our gallant young men have expressed their determination of slaying the monster, or perish in the attempt. Armed with death-dealing weapons ... little doubt is there that the days of the fear-inspiring "Yah-Hoo" are numbered.

Glenn Innes Examiner, 5 Jun 1888.

Case 78. September 1888. Jindalee, 8 km north of Cootamundra, NSW

"THE YAHOO!

... it is reported that a gorilla or yahoo has been seen several times in the neighbourhood of Jindalee and even close to town; and it is said that his unearthly cooey has been heard at night.

... it is to be hoped that our gendarmes will "run him in" ...

Cootamundra Herald, 22 Sept 1888.

Case 79. Mid to late 1800s. Brindabella Mountains, NSW/ ACT border

In 1891 a traveller mentioned the region's Hairy Man tradition:

"There are some curious additions afloat in the mountainous country in which we travelled. Old drovers talked of having seen a hairy man about 9ft. high, at whose approach the dogs ran away yelping. No person has ever been able to accurately describe this individual, but the belief in his existence still lingers around Uriarra and Brindabella.

Sydney Morning Herald, 20 Feb 1891.

Case 80. Late 1880s. Einasleigh River, north QLD

Arthur Bicknell

An old settler on the Einasleigh (a tributary of the Gilbert) told adventurer Arthur Bicknell that he'd seen a strange, bipedal animal prowling around. It was as tall as a man, had long arms and huge hands, and made a strange moaning noise. He suspected that the "wood devil", as he called it, had killed several of his dogs.

That evening Bicknell, armed with a revolver, climbed a tree overlooking the place the animal had been seen:

> I had not long been in the tree when I heard the peculiar moaning noise … and also the crashing and breaking of the brushwood … it was nearly dark and the moon only just rising. I could see nothing. The noise had an unearthly ring about it. I am not easily frightened, but … slipping down from my perch … I waited a moment to fire two or three shots … in the direction the sounds came from, and then turned and bolted for the house. If the devil himself was after me I could not have made better time.
>
> [Next morning] When the old man appeared, we [went] to the place … There sure enough he lay, as dead as any stone, shot through the heart … He was nothing but a big monkey, one of the largest I have ever seen, with long arms and big hands … These huge monkeys or apes are common in Nicaragua, but this one was certainly the largest I ever came across.

Arthur C. Bicknell, *Travel and Adventure in Northern Queensland*, Longmans, Green and Co., London and New York, 1895, pp. 173-178.

Strangely, Bicknell didn't seem to think it odd that an exceptionally large Central American monkey, presumably a Spider Monkey, should have found its way to the wilds of northern Australia. Spider Monkeys, in any case, grow no taller than 3 feet 7 inches, don't kill dogs, are arboreal and almost always quadrupedal.

"The wood devil" – an illustration by J.B. Clark based on sketches by Arthur Bicknell

In J.B. Clark's fanciful illustration, based on a sketch by Bicknell, the "wood devil" is man-sized and has a short tail. Bicknell's failure to preserve even the skull of such a remarkable animal seems highly suspicious.

Case 81. 1890 or '91. The Crowther Mountains, 10 km west of Koorawatha, NSW

In an article dealing mostly with a sighting in the same area in 1893 (Case 90) the *Maitland Mercury* mentioned that "two or three years ago ... a hairy man, or ourang-outang, had been seen in the Crowther mountains ..."

The Maitland Mercury and Hunter River General Advertiser, 20 Apr 1893.

Case 82. c. Early 1890s. Nimmitabel, NSW

In November 1912, a former resident of Nimmitabel, prompted by reports of a yowie sighting in that area (Case 130), recalled an earlier event:

> "Sir,
>
> … some twenty years [ago] I was … a resident of Nimitybelle, and beyond doubt something of the kind is there …
>
> A resident near the edge of vast jungle, in which the animal dwells, awoke early one spring morning to find it taking a nap, or meditating, with head on folded arms, on the garden gate. Immediately the door was opened it fled away.
>
> I heard of … this innocent and defenceless "wild man of the woods" … once killing a dog that attacked it, but who in self-defence might not do the same?
>
> "Humanity"
>
> Richmond, November 6.

Mercury, Hobart, 11 Nov 1912.

Case 83. c. Early 1890s. Bocks Hill, near Bathurst, NSW

> "Some years ago it was reported that a hairy man – supposed to be a gorilla or baboon – used to knock about the Bocks Hill and frighten people. We have not heard of him for some time now.

Bathurst Post, 15 Oct 1897.

Case 84. 1891. Blue Mountains, NSW

Although the editors of the *Bathurst Times* assured their readers that the following was "a narrative of fact" sent to them by "a well-known ex-bank manager [from] Oberon," it sounds very much like a tall story.

> "THE HAIRY MAN OF THE BLUE MOUNTAINS
>
> While ascending "precipitous rocks" in the vicinity of Kanangra Walls, the adventurous ex-banker and his friend "Hobby" came upon "the imprint of a huge foot".

We clambered on [until] when entangled in undergrowth, we were startled by the noise of timber breaking, and a low, growling, grating sound. We cautiously proceeded until … there before me stood an animal of the baboon species on perpendicular cliffs.

I breathlessly said, "Shoot him", but Hobby, now almost delirious with excitement, cried, "No, I will take him alive or die." He approached, and my mate and this hairy man confronted each other. We agreed to grasp the arms, when to our surprise the animal swung himself over the cliff by a huge vine, and descended in that manner until he disappeared in the gorge beneath.

The Bathurst Times, 9 Jul 1891.

Case 85. November 1891. Enoch Point, 6 km south of Kevington, VIC

"ALLEGED CAPTURE OF A WILD BOY
By Wire – From our Correspondent.
Mansfield, Thursday.

Information has reached here that a boy, covered with hair, his finger and toe nails 5 inches long, had been captured in a log at Darlingford, 26 miles from here … he was secured with much difficulty, and his captors now have him in an iron cage made expressly for the purpose.

A Church of England minister, arrived from Jamerson today, confirms the report … He was found at Enoch's Point, 20 miles from Jamisson. The hair on his head is over 3 feet and the hair on his body is 4 inches long. He … is full-grown, being apparently 16 years old. He tries to speak.

Age, 13 Mar 1891, p. 5; *Weekly Times* (Melbourne), 14 Mar 1891; *Gippsland Farmers' Journal and Traralgon, Heyfield and Rosedale News*, 17 Mar 1891; *Snowy River Mail and Tambo and Croajingolong Gazette*, 21 Mar 1891.

It seems very odd that the *Age* consigned such a fascinating item to page 5, and that no other paper featured it prominently.

In fact, the three other papers that ran the story seem to have lifted it almost entirely from the *Age*, apart from adding that the boy lived "on roots dug out of the ground," "is supposed to be a half-caste" and "is unintelligible."

As the boy was never said to be anything other than human, the claim that his body was covered in four-inch-long hair seems suspicious. No human being could sprout a furry pelt no matter how long they roamed the wilds. (None of the other, better-documented cases of feral men, boys and girls have featured unusually dense body hair.)

The fact that the story simply "died the death", never making its way into scientific journals, strongly suggests it was the invention of the *Age's* unnamed Mansfield correspondent.

But if he really did exist, the "wild boy" couldn't have picked a better place to hide – Enoch Point is situated at the very end of a dirt road, on the edge of a cliff overlooking some of the wildest country in Victoria.

Case 86. 29 January 1892. Near Armidale, NSW

(Because Days River appears to have had a name change, we have been unable to establish exactly where these events occurred).

"A STRANGE ADVENTURE

> A correspondent writes thus to the Armidale Express:
>
> GENTLEMEN, Please publish the enclosed account of an adventure which happened to four men while out mustering cattle on the lower end of the Kangaroo Hills run, and I may mention the account is perfectly true, and can be verified on application to the manager of the Kangaroo Hills station.
>
> W. G. C. B.
>
> On January 29th a party of four stockmen went down to the lower end of the run to muster cattle, and [camped] on the Day's River, about half a mile above the Bar. Just as it was getting dark they were fishing opposite a big, steep spur of the mountain, which ran right to the edge of the water … when they were startled by a heavy splash … right in front of them, like a large stone being thrown in ... two more splashes came … and each one thought it was the other who had done it. Presently three more splashes came in quick succession.
>
> The men began to get alarmed … One called out "Who's throwing stones over there?" There was no answer, they heard something moving on the rocks, but could see nothing, as it was now quite dark. Presently another stone fell right at the feet of one of them, splashing water all over him.
>
> They [went] back to the fire as fast as they could … began to talk matters over … when they distinctly heard the steps of some heavy two-legged creature

crossing over the gravelly bed of the river, coming towards them. They … began to prepare for a hasty flight, if necessary, and were busily engaged in strapping their swags on to their saddles, when a heavy stone, evidently thrown from a short distance, came with terrific force and struck the fire, scattering it in all directions. The men instantly seized their bridles, and ran [to] their horses … and found them snorting and in a terrified state … mounted them bareback; then … decided to go back and get their saddles if possible.

[While] at the camp … saddling their horses … more stones were thrown at them; they … galloped off … After about a mile … they stopped … when another stone fell about six yards from them … evidently thrown from a long distance. They galloped off again … and did not stop till they had gone several miles.

The country … was very mountainous, being about the roughest of the Day's River gulfs, and they ran great risk of breaking their necks riding over it at such a pace on a dark night.

On arriving at the junction of Kangaroo Hills Creek and the Day's River they stopped again, and thought … they had given the animal – whatever it was – the slip; and indeed he troubled them no more for about an hour, when the horses began to snort, and tried to break away. At this instant a stone was thrown with great force, and struck the ground in front of them, passing quite close to the head of one man. They galloped off again, and crossing the river rode up past Thunderbolt's Cave, and up a steep spur … When they had nearly reached the top, their horses began to get exhausted, and they were compelled to stop.

They remained there for some hours, and, just as daylight was approaching, thought they were at last in safety and were preparing … to sleep … when presently one of the men distinctly saw the form of a large creature, resembling a man, being about the same height, but much larger in the body, standing about 50 yards above them, on the spur they had been going up, and was directly in front of them, preventing them from going any farther.

He stood for a moment in a clear place … and could be distinctly seen against the sky, in the pale light of coming day … then moved slowly and silently down the hill. All this time the horses were … snorting as if they smelt something they were afraid of. Presently they could see the animal sneaking quietly up the hill towards them …

They galloped off again down the spur. There were no stones thrown till they were in motion, when several flew swiftly past them, and they narrowly escaped being hit … The animal followed them for a short distance, and then, after throwing one more stone, made off up a very steep spur, a place no horse could possibly climb, and they saw no more of him.

It is reported a gorilla was seen about three years ago, on Guy Fawkes River, by a man, who fired three shots at him without effect. The four unhappy men who were chased about all night, in this singular manner, are all quiet, reliable men … and no-one here doubts their story, as they … are prepared to swear this account is true in every particular …

The Maitland Mercury and Hunter River General Advertiser (quoting the *Armidale Express*), 11 and 16 Feb 1892.

So, once again we have a cranky yowie harassing fishermen – a scenario that will recur in the modern era. (e.g., *The Yowie*, pp. 165-67)

Rocks narrowly missing witnesses are frequently mentioned in yowie reports. We know of no one being hit by large stones, so perhaps the creatures are simply "warning people off." A few witnesses have been hit, very lightly, by small stones. (e.g., *The Yowie*, p. 291)

Case 87. August 1892. Near Pyramul, NSW

"The old story of there being some creature like a huge ape at large among the mountains in our district has again revived. Several respectable farmers, whose word is beyond question, aver that on more than one occasion some years ago both themselves and members of their families caught a glimpse of the animal near their farms on the Pyramul and Crudine Creek.

Being "chaffed" by the incredulous, they became very reticent … but when interviewed seriously, their accounts tallied in respect to the appearance and mode of progression of the unknown. They all agree … that the creature is somewhat like a large ape covered with a brownish hair, and hobbles along clumsily on his two feet.

A party of young men armed themselves and camped out for a couple of weeks a few years ago, but did not succeed in finding it, although shortly afterwards a lad who was out cattle-hunting saw it quite close, on a rocky mountain … It is very probable that it is an escaped ape, which must have got away from his keeper years ago, and lived in the bush ever since.

Sydney Mail and New South Wales Advertiser, 20 Aug 1892.

For other reports from the same area, see Cases 20 and 42.

Case 88. September 1892. Bungonia, NSW

NOTES FROM BUNGUNIA. (BY "FLY.")

> … "the hairy man" is reported as being close to town and frightening numerous people on the Spring Pond and Jerrara Creeks. A party will shortly be organised to attempt his capture. Let us hope they will be fortunate enough to slaughter this truly ferocious animal.

Goulburn Evening Penny Post, 15 Sept 1892.

Yowies still haunt the Bungonia area. (e.g., Cases 244 and 263)

Case 89. November 1892. Deepwater, 25 km east of Emmaville, NSW

It is reported that some animal resembling a gorilla has been seen near Deepwater.

The Maitland Mercury and Hunter River General Advertiser, 12 Nov 1892.

Case 90. April 1893. The Crowther Mountains, 10 km west of Koorawatha, NSW

The sensation created two or three years ago by the report that a hairy man, or ourang-outang, had been seen in the Crowther mountains [Case 81] was revived the other day by … Mr. Costello, of the Koorawatha Hotel, that … he came upon the self-same curiosity … not more than 30 yards distant.

Unfortunately, this was another instance of what a man may see when he is without a gun. The morning was perfectly clear, so that there could have been no mistake; indeed, Mr. Costello is prepared to verify his statement by statutory declaration. It is said that some years ago an ourang-outang was lost from a menagerie while travelling from Cowra to Young, but whether this explanation is sufficient to account for what Mr. Costello saw … we cannot say.

The Maitland Mercury and Hunter River General Advertiser, 20 Apr 1893.

In 1979 or 1980, "a huge, black, hairy, ape-shaped creature, approximately eight feet tall" was seen where the Grenfell to Cowra Road crosses the Crowther Mountains. (*The Yowie*, p. 248).

Case 91. Late October 1893. About 5 km south of Captain's Flat, NSW

Arthur Marrin and family.

"Mr. Arthur Marrin, cordial manufacturer, met with a rather awkward reception as he was going in to Captain's Flat a few days ago with a load of cordials (says a Braidwood journal).

Shortly after getting upon the turn off from the Cooma road … he noticed his dog running up out of the bush [and] down the road in a terrible scare. He got down to see what had frightened him, when a formidable animal, with which he was entirely unacquainted, jumped up the lower bank on to the road. It frightened him quite as much as it did the dog, as it was standing up on its hind legs with its fore feet stretched out like the arms of a man. The road, being a cutting on the hill side, was narrow, and the animal was making for him, either to follow the dog or spring upon himself.

Being unarmed [he] picked up a stone which lay close … which he threw at the beast, striking it on the temple and bringing it to the ground. He then ran up and finished it with the butt end of [his] whip. [Later] he put its body in the cart and brought it home [to Braidwood].

We paid a visit to Mr Marrin's factory [the following day] and inspected it. [its torso] was four feet [1.2 m] long, 11 inches [28 cm] across the forehead and had a face very much like a polar bear. It weighed over seven stone [98 lbs or 45 kg]. Its forearms were very strong with great paws that would be capable of giving a terrible grip. It was a tan colour like a possum with strong hair on its skin. When Mr. Marrin encountered it, it stood between 6ft. and 7ft. high … [it] was a female.

Some people think it is identical with a beast which has frightened several teamsters travelling through Parker's Gap on the Cooma road … so much so that that they have left their horses and ran away. Such an animal has been reported as visiting selectors' places at Molonglo and Foxlowe and there have been reports of the presence of similar ones in the Budawang and Sassafras ranges. It has gone by the name of the hairy man.

Goulburn Evening Penny Post, 28 Oct 1893 (quoting the *Braidwood Dispatch).*

Three days later, the *Dispatch* journalists wrote that they had measured the carcase "with the assistance of Constable Burney" and claimed that the creature's hind legs were "nearly as long as a man's" before adding this revealing aside: "Of course, it may be a wombat, and it may not …"

It is highly suspicious, of course, that nobody rushed to photograph the remarkable carcase. (By 1893 there would have been a camera or two in Braidwood.) To make matters worse, the "great number of callers" who visited Marrin's property to view the carcase were disappointed, as "unfortunately he buried it on Saturday." [The day after he killed it.]

That was, the journalists lamented, "a great pity … as there are numbers of persons who put the whole thing down as a fairy tale ..."

Yeah – right.

Goulburn Evening Penny Post, 28 and 31 Oct 1893; *Maitland and Hunter River General Advertiser,* 31 Oct 1893 and several other papers.

Case 92. Late 1800s. The Jingera, south of Captains Flat, NSW

In an article about antipodean wildlife, Mr. Murray Ashton related a story about the Australian Hairy Man:

"THE HAIRY MAN

I had at Tapanui a most engaging companion – a well-known Dunedin scientist – who, our conversation chancing on natural history, asked: "Did you ever hear of the hairy man?" On my replying in the negative, he said: "I have, and what's more, I've seen him," and he proceeded to relate the following story:

"… accompanied by three friends, with plenty of arms, ammunition, a tent, and tucker … we started for the Jingera Mountains, New South Wales.

"One night we … turned in pretty early [but] were suddenly startled … by an unearthly series of sounds – something between the roar of a lion and a clap of thunder – which seemed to reverberate all over the country … we sprang to our feet and exclaimed together: "The hairy man!"

"I was armed with a Winchester, and my friends had shotguns. We … lay peering into the darkness. The cries or roars were repeated, and seemed to be drawing nearer … The minutes passed like hours, and we were all in a terrible stew. Such unearthly cries we had never heard before. Suddenly we saw pass by a shadowy, indistinct mass some 50 yds or so from where we were, and gradually its cries died away in the distance.

"There was no rest for us that night, and at daybreak we made straight for where we thought we had seen our uncanny visitor, and there, sure enough, we found his tracks. What were they like? Well, as near as possible, I should say, to the marks that would be made by a gorilla. Indeed, scientific men say that probably that is what it is. Several people have seen him, and describe him as fully 9ft high. He appears to be quite harmless to mankind beyond nearly frightening them out of their lives, and is supposed to live on roots, kangaroos, and wallaby.

"One day he was seen to enter an empty shepherd's hut and to come out with some bread which had been left on the table. On another occasion he was seen by a lonely shepherd, who was so terrified that he made straight for the run owner's mansion and broke into his employer's bedroom at 2 in the morning, demanding his cheque.

"Anyway, it is an interesting mystery which, I believe, has not been solved up to now."

Otago Witness, 23 Nov 1893.

Case 93. Late 1800s. Near Walcha, NSW

"THE AUSTRALIAN HAIRY MAN
"Queenslander" writes:

The Australian Hairy Man.

I read with considerable interest in your last issue an account by Mr. Murray Ashton of a hairy man in Australia, and as some of my friends, and no doubt some of the public also, have grave doubts as to the existence of such a being, I should like to state that I can vouch for the correctness of his paragraph.

In the New England District ... about 30 miles from Walcha, this hairy man (or another of the same kind) was seen by several people on a station there, and the manager ... got up a hunting party especially for the pursuit of his lordship. But ... they returned without any spoil, having seen not a trace of the animal.

Otago Witness (New Zealand), 30 Nov 1893.

Case 94. 3 October 1894. Between Snowball and Jinden, NSW

"THE WILD MAN OF SNOWBALL

On the 3rd of October last young Johnnie McWilliams was riding from his home at Snowball to the Jinden P.O. [Braidwood district].

When about half-way the boy was startled by the extraordinary sight of a wild man or gorilla [that] suddenly appeared from behind a tree, about thirty yards from the road, stood looking at him for a few seconds, and then turned and ran for the wooded hills a mile or so from the road. The animal ran for two hundred yards across open country before disappearing over a low hill, so that the boy had ample time to observe the beast.

The boy states that he appeared to be over six feet in height, and heavily built. He describes it as "a big man covered with long hair". It did not run very fast and tore up the dust with its nails, and in jumping a log it struck its foot against a limb, when it bellowed like a bullock. When running it kept looking back at the boy ... It was three o'clock in the afternoon, and the boy describes everything he saw minutely. [He] is a truthful and manly young fellow, well acquainted with all the known animals in the Australian bush, and insists that he could not have been mistaken.

For many years there have been tales of trappers coming across tracks of some unknown animal in the mountain wilds around Snowball. Of course, these tales were received with doubt, and put down as clever romancing on the part of the possum hunter, but the story of Johnnie McWilliams is believed by all who know the boy … The proof of the existence of such an animal in New South Wales should be of some interest to the naturalist. – Braidwood "Dispatch".

Goulburn Evening Penny Post, 1 Dec 1894.
Also *Queanbeyan Observer*, 30 Nov 1894; *Cooma Express*, 30 Nov 1894.

Case 95. Early 1895. Horton River, 20 km west of Bingara, NSW

"Several residents of Bingara, including a clergyman, have lately seen a peculiar animal known as the ya-hoo, hairy man of the woods, or gorilla, in some rough country near the Horton River. As soon as it was seen the animal rushed off into the bush, but its tracks clearly showed that it was no ordinary animal. Its appearance has caused considerable consternation in the vicinity.

Clarence and Richmond Examiner, 9 Mar 1895.

These events occurred 20 kilometres or so from where William Telfer saw a Hairy Man 12 years earlier. (Case 63)

Case 96. July 1895. Moss Vale, NSW

"… a couple of young ladies were alarmed and disturbed … by a supposed "bear" on Monday night. They were resting on the seat on the Moss Vale road when the hairy monster came up to them through the bush. It was carrying a fowl. They all scattered.

Bowral Free Press and Berrima District Intelligencer, 20 July 1895.

Case 97. July 1896. A few kilometres west of Tenterfield, NSW

> A Tenterfield resident the other evening reported that he had been attacked by a large hairy animal like a gorilla. An armed party is in pursuit.

Barrier Miner, 15 Jul 1896; *Brisbane Courier*, 17 Jul 1896; *Mercury*, 18 Jul 1896.

According to some of the later reports, the encounter occurred on the Clifton Road. A police constable led an armed party to the scene, but they returned empty-handed.

Case 98. February 1897. Near Candelo, south-east NSW

> A STRANGE ANIMAL
>
> The Wyndham correspondent of the Pambula Voice says: A very strange animal was recently seen … by a young fellow named John Wilmott … between New Station and Candelo … [he] and another young fellow (Edward Power) were walking along looking for a horse when they suddenly heard a noise like the roaring of a bull, which seemed to come from the jungle that abounds in this part of the district.
>
> [They] thought that some cattle had gone into the jungle, and Wilmott rolled a good size stone down the hill, when to his surprise the strange animal proceeded from the thickest of the scrub. It passed over a small clear patch … and Wilmott states he saw it plainly, he does not know whether to call it a man or a beast, being about 5 feet in height, it had arms like a man which appeared to be of great length. Its head was rather small and round something like a monkey's. He said that several residents have seen it at various times.

Delegate Argus and Border Post, 25 Feb 1897.

Case 99. Late 1800s. Quidong District, 20 km west of Bombala, NSW

Letter to *The Sydney Stock and Station Journal*, 1 Sept 1916:

> "THE GORILLA
>
> The photo enclosed is of a place called "The Bluff" … where the gorilla, which was seen on Monaro, is said to have lived.
>
> A man was camped down there to shoot kangaroos. He was skinning a kangaroo one day, when he looked up and saw coming towards him a huge hairy animal. It had a large body with long, powerful arms. It had a face something like a man's – only it was covered with hair.
>
> [He] jumped on his horse and galloped away as hard as his horse could go; but the gorilla followed him for a good way, pelting huge stones at [him] all the time.
>
> Billy (that was the man's name), galloped all the way here, never stopping once to give the horse breath. When he got here, he told father about the gorilla and borrowed one of our horses to ride to Delegate. He got a party of six men together and they all rode, with loaded rifles, to look for the gorilla. However, they couldn't find any traces of it.
>
> Yours, etc.,
> "Michaelmas Daisy"
> Delegate.

Case 100. Late 1800s or early 1900s. Currickbilly Range, NSW

Few people knew the mountains and gorges of south-east NSW as well as the surveyor Charles Harper (1840-1930) who worked throughout the region for about 50 years. On hearing of the Summerell incident (Case 130), he wrote to the Sydney *Sun* to describe his own close encounter with one of the creatures:

> "In various parts of the southern district …on the coastal slopes, … over a long period, I have met men (and reliable men at that) who unhesitatingly assert that they had seen this hairy man-shaped animal at short distances. They were so terrified at the apparition and the hideous noise it made that they left their work

as timber-getters, and at once cleared out … leaving their tools and work behind them. At the risk of being considered … the reincarnation of Ananias or the late Thomas Pepper, I will describe this animal once seen as briefly as possible.

Charles Harper

I had to proceed some distance into the heart of these jungles … accompanied by two others, and two large kangaroo dogs … On the [second] night, about 9pm ... we heard a most unusual sound, similar to the beating of a badly-tuned drum, accompanied by a low, rumbling growl. The dogs were supposed to be able to tackle anything. But … they seemed utterly demoralised; they would not bark, but whined, and made to come into the tents.

The horrible sounds gradually drew nearer and our thoughts flew to escaped tigers ... We had no firearms, only a scrubhook and an axe ... one of my companions [placed] a large bundle of leaves and dry kindling on the smouldering camp fire … they flickered up into a big blaze ... when a most blood-curdling sight met our gaze.

A huge man-like animal stood erect not twenty yards from the fire, growling, grimacing, and thumping his breast with his huge hand-like paws. I looked round and saw one of my companions had fainted. He remained unconscious for some hours. The creature stood in one position for some time, sufficiently long to enable me to photograph him on my brain.

… its height … would be 5ft. 8in. to 5ft. 10in. Its body, legs, and arms were covered with long, brownish-red hair, which shook with every quivering movement of its body. The hair on its shoulder and back parts appeared in the subdued light of the fire to be jet black, and long; but what struck me as most extraordinary was the apparently human shape, but still so very different.

I will commence its detailed description with the feet, which only occasionally I could get a glimpse of. I saw that the metarsal bones were very short, much shorter than in the genus homo, but the phalanges were extremely long, indicating great grasping power …The fibula … was much shorter than in man. The femur … was very long, out of all proportion to the rest of the leg.

The body frame was enormous, indicating immense strength … The arms and forepaws were extremely long and large, and very muscular, being covered with shorter hair. The head and face were very small, but very human. The eyes were large, dark and piercing, deeply set. A most horrible mouth was ornamented with two large and long canine teeth. When the jaws were closed they protruded over the lower lip. The stomach seemed like a sack hanging halfway down the thighs, whether natural or prolapsus I could not tell. All this observation occupied a few minutes while the creature stood erect, as if the firelight had paralysed him.

After a few more growls, and thumping his breast, he made off, the first few yards erect, then at a faster gait on all fours through the low scrub. Nothing would induce my companions to continue the trip, at which I was rather pleased than otherwise, and returned as quickly as possible out of the reach of Australian gorillas, rare as they are.

The Sun (Sydney), 10 Nov 1912.

The "Bombala Anthropoid", as depicted by newspaper artist Will Donald in 1912.

Although the end result looks rather comical, it is evident the *Sun's* artist, Will Donald, was genuinely attempting to sketch the creature as described by Harper.

Amid the wealth of detail in Harper's report, his description of the creature's grotesque, sagging stomach has generated most comment. Interestingly, a similar titanic tummy was seen many years later: in 2001, on Glen Street, Woodford, right on the edge of the Blue Mountains wilderness. At 4 o'clock one April morning, as "D", a baker, was about to leave for work, he saw a seven-foot-tall ape lurking in his driveway. In addition to having all the characteristics of the average yowie-in-the-street, it sported "a huge stomach … a gut [that] hung … like a tucked-over-the-pants beer belly." (*The Yowie*, p. 288.)

Although several other people have reported sounds suggestive of chest thumping, Harper is the only witness we know of who actually observed it. His assertion that the creature's shin was much shorter than a human's and that the thigh was "very long, out of all proportion" matches the observations of some modern-era witnesses. Interestingly, even his assertion that one of his companions fainted and remained unconscious for some hours, has been reported elsewhere.

Case 101. c. late 1800s. Hawkesbury River district, NSW

On reading of Charles Harper's sighting (Case 100, above), Mr. Horace Saxon, of Sackville, Hawkesbury River, evidently a pompous know-all, wrote to the *Sydney Sun* to question the veteran surveyor's powers of observation and to convey some dubious information supposedly given him by witnesses in the Hawksbury district.

While one detail – that most sightings involve only a solitary, apparently male, creature – agrees with the data in our files, most of what he says sounds like utter rot.

As the full text of his letter is teeth-grindingly tedious to read, we offer here a heavily-edited version:

> "Mr Harper's account ... will cause no surprise to a number of old hands, who have long kept [quiet about] what they have seen or heard .... In this ... district there are at least five persons who have seen the creature ...
>
> If Mr. Harper had more skill as a naturalist, or a longer time to observe the creature, he would have seen that it is neither ape, nor man, but may be best described as marsupial man. The so-called hanging stomach ... was in reality the pouch in which it carries its young, like other marsupials.
>
> Only one mature creature has been seen at a time [so] There is no reason to suppose that there is a tribe ... but three of the five observers noted that he – for it is always a male – carried in a pouch a black baby. The black baby, when he pokes his head out ... is strikingly like an aboriginal child, and much more human than its grown parent.
>
> Now this fact of a male marsupial having a pouch, to no doubt assist the female in carrying the heavy young, is such a novelty to scientists, and was sure of such a howl of ridicule that the discoverers have wisely kept [quiet].
>
> ... the creature has only been seen or heard at intervals of several years ... [always] near the coast range of mountains, and though at points as far separated as Gippsland and North Queensland, it is quite possible that there is only one pair with their young. Probably the creature wanders up and down the coastal range, moving according to the season of the year, or as its food supply varies.
>
> Two of our local observers had the forethought to take plaster casts of the creature's tracks; but have not shown them to scoffers.

Sydney *Sun*, 17 Nov 1912.

Case 102. December 1898 – March 1899. Upper Colo, NSW

"The "hairy man" … has made his appearance at Upper Colo … playing havoc among the quinces and apples along the river. It was first seen by Charlie Blundell, whilst … passing through his orchard in search of flying foxes, which he thought had been destroying his early peaches and apricots.

The Windsor, Richmond, and Hawkesbury Advertiser, 24 Mar 1899.

Case 103. May-June 1899. Near Wandsworth, 25 km north-west of Guyra, NSW

"A WILD MAN DISCOVERED, LIKE A GORILLA

One of the Inverell papers reports that there is considerable perturbation … in the neighbourhood of Wandsworth over the appearance of a gorilla in that vicinity.

About a month ago a man named Meaney saw a large animal of some sort in the mountain between Wandsworth and Paradise Heads [Paradise West?] and on Monday last (June 5) a young man passing by the same way also caught sight – or says he did – of a similar kind, i.e., a tall, hairy, ferocious, gorilla-looking creature.

On Tuesday about 20 residents of Wandsworth, mounted, and armed to the ears, set out in search of this strange denizen of the bush … If they do happen to capture such an animal … it will authenticate all the stories that have from time to time been told of the existence of the ourang-outang in the wilds of Australia.

Singleton Argus, 15 Jun 1899; *Albury Banner and Wodonga Express*, 16 Jun 1899.

Case 104. c. 1900. Tinderry Mountains, 15 km east of Michelago, NSW

Denis Debenham recalled that his grandmother, Mrs. Tyrie, told of seeing a "gorilla" at her property, "Little Tinderry", when she was a young woman. She watched the animal walk on its hind legs down to a creek and stoop to drink. The experience so impressed her she that she spoke about it for the rest of her life.

Denis Debenham, interview with Tony Healy, 1977.

CHAPTER TWO
THE EARLY MODERN ERA, 1901–1974

Case 105. 1901. Brindabella Ranges, NSW/ACT

In 1903 Mr H.F. Cox told his friend John Gale about something that occurred two years earlier.

"Mr Cox was ... camped alone in the ranges ... enjoying his billy of tea in the afternoon, when his attention was drawn to an enraged cry, between a howl and a yell, in the thick scrub of a gully close by. He instantly seized his rifle and looked in [that] direction ... There he saw a huge animal in an erect posture tearing through the undergrowth, and ... it was out of sight before he could bring his rifle to his shoulder.

He distinctly heard the crashing of the undergrowth in its flight, and he followed [but its] speed was greater ... its howling and yelling continued.

That it was no creation of an excited imagination – (and from what I know of Mr Cox, he is ... a remarkably cool, intrepid fellow, too well enlightened and educated to magnify a simple fact into a chimera) – is confirmed by this, that in his pursuit he met several wallabies tearing up the gully in such alarm that, though passing close by, they took not the least notice of him. These were followed presently by a herd of cattle similarly scared.

John Gale, *An Alpine Excursion*. Queanbeyan, 1903, pp. 85 – 89.

Case 106. c. 1901. Nymboida district. Aboriginal lore

"Sir,

With reference to the aborigines at Nymboida, I beg to report ...

MYNGAWIN – A ghost is supposed to be a wild hairy-man wandering about the bush on the tops of the mountains, but is perfectly harmless and only frightens people.

Yours, etc.,

F. Brown.

Letter to the editor, *Science of Man*, 1 Jun 1901.

Case 107. 1901 or 1902. Coorambeen Creek, near Jervis Bay, NSW

In January 1977, 86-year-old Henry Methven told Patricia Riggs of the *Macleay Argus* about a little creature he'd seen while hunting near Jervis Bay in about 1901. Having become separated from his companions, the then 10-year-old returned alone to a temporary camp.

> "I was stripping off my shirt and when I looked around, the Hairy Man was standing right behind me. He was only about ... two or three foot ... a handsome little fellow... he had a long straight nose and he was the colour of a real full blood ... dark and coppery ... everything about the little bloke ... seemed to be human."
>
> The creature was strongly built with a short neck. There was hair on the back of its hands. On its head, the hair was about two or three inches long and "a bit smoky looking, a bit grey." But its body hair was "darkish brown". That was as much detail as the startled boy could absorb: "I took off into the bush and got stung with stinging nettles.
>
> "The next day we tracked him. He had feet like a human's ... five toes." On a nearby ridge, they found evidence that the little creature lived with others in a small cave and dined on shellfish. Henry said all the tribal elders knew about the creatures. They called them *wallathegah*. They were said to be harmless, but to have a great fondness for honey. Earlier that day, Henry's party had harvested honey from a native beehive and he had carried it back to camp in a coolamon. The elders said, "He could smell the honey and he followed you along."

Macleay Argus (Kempsey), 6 Jan 1977. Credit: Patricia Riggs.

Henry's *wallathegah* was evidently one of the small creatures known to other Aboriginal language groups as *junjudees*. It is worth noting, also, that in July 1985 a non-Aboriginal family saw a much larger Hairy Man just a couple of kilometres from the site of his encounter. (*The Yowie*, p. 255)

Case 108. 2 November 1902. Near Moe, VIC

"A STRANGE ANIMAL

On Sunday evening .... Mr Cosgriff's farm ... 3 miles from Moe, was alarmed by the appearance of a strange looking animal helping itself to some skim-milk. Miss Cosgriff, who first noticed the animal, raised the alarm, and ... members of the family arriving on the scene, summoned the dogs, who succeeded in frightening the animal, which made off towards some undergrowth not very far distant.

The dogs followed ... but he ... shook them off as if he were tossing bones into the air. When the party followed up ... the animal had made good his escape, and it is thought made across the Latrobe River to Spike Island.

No conclusion could be arrived at as to [its species]. It is described as being something like a huge bear, and has been heard making its way to the respective farms in the neighbourhood. Its tramp, it is said, resembling that of a baby elephant.

On Thursday evening last, Mr Alex. M'Lean and Mr Bert. Savige (accompanied by Willie Ellis), drove out to Mr Cosgriff's farm, armed to the teeth ... The night was dark, but the animal nevertheless put in an appearance. Willie Ellis was left in charge of the horses. Two shots were fired, but as the moon was obscured, and the rain was falling in torrents, true aim was out of the question. The animal in its exit, passed the lad Ellis, and what followed can only be imagined, as the latter let the horses "slide," and they galloped towards Cosgriffs house ...

When the animal was first seen at Peel's, it caused equal consternation as at Cosgriff's. It is said that it dashed through the gate at the latter homestead, wrenching the hinges off ... It is surmised that [it] may be a bear or some other animal that might have effected his liberty from some passing show.

On Saturday evening "Jim" Cosgriff had another encounter with the strange animal, as he was on his way home from Moe. Being unarmed he could only beat him off his greyhound with a stick, [then] the monster made his way through the undergrowth towards the river.

On Sunday night a number of townsmen ... Rifles, revolvers and other weapons ... were in readiness ... but there was no sign of the creature. – "Advocate."

Morwell Advertiser, 21 Nov 1902.

Although the critter was described as looking "something like a huge bear", it sounds decidedly yowie-like to us. Whereas bears move quietly, the yowie's heavy tramp is often mentioned.

On the other hand, another creature thought to be a bear was seen near Darnum, just 20 kilometres west of Moe in 1933. (Case 159)

Case 109. Early 1900s. Burnt Bridge, Kempsey, NSW

Mrs Mamie Mason was born into the Aboriginal community at Burnt Bridge, about four kilometres south-west of Kempsey. Her grandparents often mentioned the Hairy Men, and said that they used to leave out honey and other food to keep them away from their camps.

Sometime in the early 1900s, two-year-old Chris Davis was carried off by a yowie. Like all her contemporaries, Mrs. Mason knew the story well:

> "On this day the little boy's mother stopped to have a drink at the creek. It was just on dark and next thing ... he was gone. They listened. They could hear him screaming. They crossed over the gullies and ... saw the Hairy Man had Chris ... it was going to the cave ... but the father got to the child. The Father pulled the child out of its arms. That was Old Man Davis, the father. The child was almost insane, they said, when they got it. You don't make up those things. You never forget them ..."

Macleay Argus (Kempsey), 2 Oct 1976. Credit: Patricia Riggs.

Burnt Bridge is about 6 kilometres east of where a Hairy Man was seen in 1871. (Case 34)

Case 110. 1902. Blue Mountains, NSW. Aboriginal lore

In an article titled "Aboriginal Traditions", the writer, "Bordad", relates some Hairy Man lore that he heard from an elderly Indigenous man.

> My informant gave me some facts concerning the "Yahoo," I cannot call to mind the aboriginal name for this animal. The "Yahoo," (as we all know) is an animal said to resemble a man only that his body is covered with long hair, and his feet are turned backwards, the toes being where the heel should be. The aboriginals really believe that such an animal exists and they are all afraid of it. My informant confidently believes that one is still living. He, indeed, offered to take me to the place where I could see it for myself. He says this strange creature is to be

seen at the Devil's Hole, a point about two miles from Katoomba. He describes this particular Yahoo as being large and strong …

A story is told by the blacks that on one occasion an aboriginal caught a Yahoo woman and took her to wife. Children were born and reared, but after a time the tribe quarrelled over the strangers and killed both mother and children.

Bathurst Free Press and Mining Journal, 19 Jun 1902.

Case 111. January 1903. The Rock, 32 km south-west of Wagga Wagga, NSW

It is rumoured that a wild hairy animal resembling a gorilla has been seen in the vicinity of The Rock Hill.

Wagga Wagga Express, 24 Jan 1903.

Case 112. August 1903. Between Wyndham and Candelo, south-east NSW

It is reported (says the 'Voice') that a gorilla was seen last week on a mountain cutting between Wyndham and Candelo.

Delegate Argus and Border Post, 8 Aug 1903.

More details appeared in the same newspaper on 15 August (quoting the *Eden Propeller*):

A great sensation was caused here [Eden] last Thursday when a gentleman came into the township and stated that he was startled by seeing what he termed a gorilla between Candelo and here. He says it was … the size of a full-grown man, with abnormally long arms and large head. It bounded right onto the cutting, gazed at him in a weird sort of way, and made a most unearthly noise … leaped over a fence and made for the ranges with the "speed of an antelope."

This happened in the general area of John Wilmott's February 1897 encounter (Case 98). Egan Peaks/The Jingera, believed by Aborigines to be an abode of the Hairy Man, is nearby, as is the spot where, in December 1977, Mr. Kos Guines shot at "a huge, black creature like a gorilla." (*The Yowie*, p. 65)

Case 113. 1904. Egan Peaks or The Jingera, near Pambula, NSW. Aboriginal lore

Researcher Graham Joyner discovered this entry in a 1904 issue of *Science of Man*:

> "Jingara. A huge mountain, supposed to be haunted by a hairy man or Yahoo."
>
> "Aboriginal Dialects: Cooma Sub-District", *Science of Man*, 23 Aug 1904, vol. 7, no. 7, p.104.

Egan Peaks, or The Jingera – said to be an abode of the Hairy Man

The experience of white settlers in the vicinity of Egan Peaks/The Jingera lends support to the Aboriginal tradition. The area has featured in several eyewitness reports from colonial times through to the modern era.

Coincidentally, several sightings have occurred in another area called The Jingera – a section of the Great Dividing Range to the south of Captains Flat. (e.g., Cases 35 and 192)

It is also interesting to note that, nearly 3,000 kilometres to the west, some Indigenous people used terms quite similar to *jingera* while discussing the Hairy Men. Near Lake Grace in south-west Western Australia, for instance, the ape-men were known as *jimbars, jimbras* or *jingra.* (Case 23)

Case 114. 1904. Byron Bay, NSW

Mrs. Sarah Ratcliff said that her father, Patrick Maher, encountered a yowie while riding home to Tyagarah from Byron Bay in 1904. At about 10.30 pm, after crossing Belongil Creek, the horse became nervous and, looking back, Mr. Maher saw a creature "like a big hairy man with no neck," standing on a heap of stones. It was seven or eight feet tall. The horse bolted and the creature gave chase, but Mr. Maher lost sight of it when he was thrown from the saddle.

He ran into a cane plantation, stayed there all night, and was still badly shaken when he reached home in the morning. The property was sold as quickly as possible and the family moved to Tumbulgum.

About a year later, three brothers who lived nearby were riding home from a dance at Byron Bay when they were pursued by a beast that leapt onto the rump of one of the ponies. It hung onto the saddle until the party galloped off the road into the home track. The rider's coat was torn and the horse was so badly injured it had to be destroyed. The injured boy was put into hospital, and according to the report, all the brothers "were … ill after their fright."

Daily News (Tweed Heads), 20 Jun 1982; Jim Brokenshire, "The Brunswick, Another River and its People," Brunswick Valley Historical Society, 1988, Page 176.

Case 115. 1904 or '05. Near Torbanlea, QLD

This odd, but very intriguing story is one of the very few that features amphibious yowies or – more likely, in this case, given their diminutive stature – junjudees.

As the creatures were amphibious, the incident was categorised, by the *Wide Bay News*, as a bunyip event.

> "THE BUNYIP AGAIN
>
> Mr. E. Williams resides on his selection, "Greenfield", about eight miles from Torbanlea and twelve miles from Maryborough, [where] there are many large lagoons … of a depth in places of 60 to 70 feet.

Four or five years ago this winter, Mrs. Williams and her [18-year-old] daughter saw two animals about three-and-a-half feet high, and a smaller one, come out of the water hole nearest the house and run about 100 yards towards them, and then run back and dive under the water. They called Mr. Williams, who was just in time to get an indistinct glimpse of one of the forms under water. The animals resembled little men covered with longish black hair.

About a year later, Mr. Williams' three sons, aged 19, 17 and 16 years ... reported seeing an exactly similar subject erect upon the bank of the lagoon. [When disturbed] it plunged into the water and swam about near the bank, distinctly visible for a short time.

There were [later reports] of others seeing something similar in the Brushwood Lagoon a few miles away, during the drought.

Bundaberg Mail and Burnett Advertiser (quoting the *Wide Bay News*), 10 Sept 1909. Credit; Tohby Riddle.

The Torbanlea – Howard area, in which the events supposedly occurred, has featured heavily in modern-era yowie reports (Cases 220 and 235) and the lagoons in question appear to drain into the Burrum River – in which a "sea serpent" was reported in 1995. (Malcolm Smith, *Bunyips and Bigfoots*, p. 63)

Case 116. 1904-05. Narrogin, WA

On 1 February 1905 this rather curious letter appeared in the *Narrogin Advocate and Southern Districts Courier*:

"Sir – for some time past an animal new to natural history has been seen in the vicinity of Narrogin, and has been a source of much speculation to all who have seen it.

It is about the height of an ordinary man, and walks upright. The head is large, with very large ears (like those of the gorilla), and nasal protuberance, and similar arms.

In winter the head and face is thickly covered with hair, and in summer it sheds the hair on the lower part of the face. During the summer it is mostly seen in shady places, sitting with its back against a tree. Some who have got close ... tell me the hair is rubbed off the back.

In winter it seems to roam more, and is found mostly along the roads, especially where there are any holes. It appears to get some kind of food from them. From observations which have been made its food is the same as that of the baboon. When it is cold and it finds a fire anywhere, it will sit for hours beside it, but does not seem to have the instinct to put more fuel on it.

The oldest settlers here tell me they have never seen or heard of it before, and several of the blacks … tell me "him debil debil, come along where sun jump up." [So it] seems to have its lair to the eastward of Narrogin.

Hoping your readers may be able to throw some light on the above.

Yours, etc.,

HAWK

Narrogin, Jan 29.

Case 117. 7 May 1905. Near Belowra, 30 km west of Bodalla, NSW

"A WILD MAN

At various times for many years past (says the 'Southern Star') different people have vowed to seeing an animal resembling a hairy man in the mountainous and almost unknown country around Beloura.

The late C. J. Byrne. who frequented these parts a good deal many years ago, is reported to have seen this strange animal, as did also a man employed by Mr Cowdroy, of Bega. The blacks in those days also spoke of such a thing, and were always very frightened when in the vicinity of what is known as Wyoila Creek. Now … Messrs J. Wall, sen, Harry Staples, J. L. Wall, and W. Roberts – who are employed up there by Mr Cowdroy, have seen the monstrosity.

Three of [them] … have addressed the following letter to this paper:- "While exploring on Sunday, May 7, we saw a strange looking animal just like a hairy man, run out of some rocks and along a beaten track, leaving a foot track similar to a man's. The footmarks were about 14 Inches long and the animal appeared to be about 7ft high with tremendously long arms, reaching to below the knees. When it got about 100 yards away it stood, and did not appear to be frightened, but after a while walked away gently in the direction of the Wyoila Creek mountains. None of us tried to hinder it, as it looked too fierce. and the dogs would not go near it, crouching at our feet.

The Maffra Spectator, 25 May 1905.

Case 118. August 1905. Euchareena, 35 km west of Pyramul, NSW

IS IT A GORILLA?

MOLONG, Thursday.

Considerable commotion has been caused in the Euchareena side of this district by the circulation of a rumour … that a hairy man, or gorilla, has been seen in the depths of the bush there. Two well-known and highly reputable residents report having seen the fearsome-looking object. A youth, who unconsciously rode up in close proximity to the animal, received such a scare that he lit out for home in record time. He has no doubt whatever but that the animal is a gorilla.

For years past strange sounds have been heard in the thick scrub close to where this animal was seen, and a project is now on foot to form a search party and solve the mystery.

Evening News (Sydney), 24 Aug 1905; *Wellington Times*, 24 Aug 1905.

Additional details were supplied by the *Narromine News and Trangie Advocate* on 1 Sept 1905 and the *Muswellbrook Chronicle*, 20 Sept 1905 (both quoting the *Molong Argus*.)

THAT STRANGE ANIMAL

On Tuesday, a talk with Mr P. Horton, of Euchareena furnished information … that he saw a gorilla right enough, and that he could not be deceived as to the character of the animal, having seen several in [the Sydney zoo].

Mr Horton, snr. stated that this beast has been in the ranges for some time, its harsh cries being frequently heard. It is on record that a gorilla was known to inhabit another part of this district for some time – it was seen by several people, including Mr A. Kerr, of Orange Bide, a man of the world, who had no difficulty in recognising the beast as a gorilla. How these two hairy representatives of Africa came to locate themselves in the Orange-Molong district partakes somewhat of a mystery; the only feasible explanation is that they escaped from some travelling menagerie.

That was not the end of the story – a similar report came from the same area twelve months later. (Case 122)

Case 119. 1906. Wyan Mountain, north-east NSW. Aboriginal lore

In an article about Aboriginal customs and folklore, E.S. Sorenson stated that "Wyan Mountain, on the Richmond River, which has many caves, is the reputed home of a hairy man of gigantic proportions and ferocious nature."

Clarence and Richmond Examiner (Grafton), 5 Jun 1906.

Case 120. June 1906. The Gulf, 20 km north of Emmaville NSW

"THE YAHOO AGAIN

> A correspondent of the *Glen Innes Guardian* ... says that Mr. E. J. Clifford has had this exciting experience about five miles north of The Gulf: He was crawling through dense scrub, when a wild animal rose up in front of him, and, as if frightened, paused for a moment. Clifford, who had a .32 Winchester [fired as] speedily as possible ...
>
> The animal which, he says, appeared to be 8ft high and 3ft broad, black in colour, and covered with hair of great length, gave several unearthly yells ... and made off. Clifford fired two more shots at it as it fled over a flat. Following it for about a quarter of a mile, he saw blood and huge foot-prints where the animal crossed a creek in the sand. The head of the animal seemed to be shaped like that of a dog. [It] at first stood upright like a human being, and ran on two legs at a great pace. It was some forty-two yards away from Clifford when first seen. The time was about midday, and Clifford saw it clearly.
>
> A rifle party of about 30 has left The Gulf in search of the creature. Clifford is said to have shown some of the miners the blood and tracks on the sand.

Richmond River Herald and Northern Districts Advertiser, 22 Jun 1906.

Case 121. August 1906. Near Eden, south-east NSW

“Mr. Alf. Smith, of the Lakes, (Eden district) distinctly saw a hairy man a short distance from him a few days ago, and … shot at it with a shot gun without effect. He says it strongly resembles a gorilla, is between 5 and 6 ft. high, has long hair over the main body, and short arms.

The Bega Budget, 29 Aug 1906; *The Southern Star*, 29 Aug 1906.

In December 1977 Kos Guines of Frankston, Victoria, encountered a similar creature about 20 kilometres north-west of Eden. He, too, fired his shotgun into its broad back, but, like Mr. Smith's creature, it didn't even flinch. (*Out of the Shadows*, p. 113 and *The Yowie*, p. 65)

Case 122. August 1906. Pyramul, NSW

“A TOWNSHIP IN TERROR.
IS IT A STRAY BABOON?

A telegram from Mudgee states that the residents of Pyramul, about 40 miles from Mudgee, have been terrorised by the appearance there of a strange animal, resembling a baboon. Several reputable people aver that they have encountered the animal, and that its appearance is such as to inspire the greatest dread.

The Advertiser (Adelaide), 22 Aug 1906; *Gippsland Times*, 23 Aug 1906.

As we have seen, there was a similar scare at Euchareena, 35 kilometres west of Pyramul, just twelve months earlier. (Case 118)

Case 123. Dec 1906. Cowan, NSW

“It would appear that Borneo is not the only place in the world to own a wild man. If three ladies who were until recently staying on a house-boat at Cowan were not greatly deceived, they saw the strange figure of a hairy man, tanned to a coppery hue, flitting about among trees and rocks with the rapidity of a will o' the wisp. They state that he climbed trees with almost the agility of a monkey, and could run … with great speed. Police from the surrounding districts toured the mountainous country, but without finding any trace of this wild being. At one place, where there is a sheer drop of a hundred feet or more, footprints were discernible down the face of what would appear to be an inaccessible cliff. If there is a man roaming about, he must have a wonderful knowledge of the hiding-places, for the police have not been able to catch a single glimpse of him.

The *Daily Telegraph* (Sydney), 3 Dec 1906.

Case 124. 1907. Near Creewah, NSW

After reading about George Summerell's October 1912 encounter with a Hairy Man (Case 130) R.W. Dawson of Goulburn told of seeing a similar creature in the same area five years earlier.

While pursuing his horse, which had run off after becoming unaccountably nervous, he saw a dark object he took to be a peculiar-looking tree stump:

> … but before I could give a second thought it bounded off at right angles. My surprise cannot be described, but I took it to be a man running to block my horse … I noticed the pace, the stride, and the force of the footfalls of my "friend." I saw it run quite 100 yards at a distance of about 75 yards away. It appeared to be a tall man with a fur cap and a pea jacket, and running with a peculiar attitude.
>
> I got my horse about two miles on, and then the facts began to force themselves on me, and I concluded that I had seen something out of the common, either a wild man or animal.
>
> [That evening] one of the Summerells informed me that he had actually seen a hairy man in the neighbourhood a short time before. His description … coincided with what I saw. I [also] spoke to Mr. Fred Heffernan, of Thoko [five kilometres north of Creewah] and he said he had had many a hunt for some animal because of the peculiar footprints he had seen …

Goulburn Evening Penny Post, 26 Oct 1912.

Case 125. August 1907. Waterfall, NSW

> Recently a lion roamed in our bush – whose career happily was brought to a finish, but now a fairly reliable rumour has it that another strange creature is at

large out Waterfall way. The animal is said to be like most intelligent of creatures, the human kind, and on this it is concluded that [it] is a gorilla. Mr. Jury, of Salt Water Creek is said to have encountered it towards evening, but owing to the increasing darkness was unable to fulfil the work of identification.

South Coast Times and Wollongong Argus, 13 Aug 1907.

This is just one of many references to a huge, hairy creature being seen in the Waterfall-Sutherland area between about 1860 and 1910 (Case 25). In fact, the critter, known then as "the wild man of the bush" or simply as "The Thing", might still be around – in 2000, two young men came within six metres of a huge, bipedal ape-man in nearby Royal National Park. (*The Yowie*, pp. 26 and 206)

The mention of an out-of-place big cat being seen in the vicinity of the "gorilla" is also interesting in light of the many "black panthers" that have been reported in yowie hot spots in the modern era. (e.g., Cases 265, 266 and 267)

Case 126. September-October 1908. Cordeaux River, NSW

"A STRANGE ANIMAL

During the past few weeks settlers on the Cordeaux River have been much alarmed by a strange animal, said to be either a gorilla or bear. It uttered loud cries at night. Some people have not ventured out at night in consequence. Men have gone to work in the bush armed, as a precaution against the unknown beast.

Evening News (Sydney), 30 Oct 1908.

The Cordeaux River has featured in other reports. Mr. B. Rixon saw a gorilla-like creature there in about 1865 (Case 29) and on 22 July 2002 a Sydney man encountered a foul-smelling, eight-to-nine-foot, hair-covered biped near Cordeaux Dam. He assumed it was "some sort of escaped gorilla or bear." (*The Yowie*, p. 289)

Case 127. June 1909. Maitland Bar, 20 km south of Mudgee, NSW

> During the past few weeks the residents of the "Bar" have been disturbed ... by noises, resembling at times a person choking, and at others a woman screaming and then crying.
>
> These ... remained a mystery till Thursday last, when at about 5pm they were again heard, and ... several persons ... were astonished to see a peculiar animal, five feet high, standing on his two legs, and at the same time brushing away with his claw-like hands the long unkempt-looking hair from his eyes.
>
> The animal is covered with long white hair and when seen was uttering the cries which have been disturbing the peace of the neighbourhood. The hairy man, or whatever he is, was only seen for a minute, and disappeared as suddenly as he came in sight.

Mudgee Guardian, June 10, 22 and 24, 1909; *Robertson Advocate,* undated clipping (probably June) 1909.

A few days after the above story appeared, the *Mudgee Guardian* reported that police believed the sightings were the work of a prankster dressed in a goatskin. The supposed prankster and his costume were never tracked down.

That, however, wasn't the end of the matter, because on 19 June the *Advertiser* (Adelaide) ran the following story:

> THE WILD MAN AGAIN
> TWO BOYS FOLLOWED
>
> Sydney, June I5.
>
> The scare at Maitland [Bar] has taken another form. Whilst two youths named Somerfield and Lue were among the hills lopping kurrajong for cattle, they heard a peculiar noise, and looking upwards ... saw what they described as a hairy man, sitting on a rock, staring at them.
>
> Taking fright, they ran and were pursued by the stranger. In his efforts to escape Lue fell into the river, being nearly drowned before Somerfield rescued him, after which they continued their flight ...
>
> Mrs. Albert Brennan, hearing cries, hastened to render assistance, but the strange man still approached with a lumbering gait, and she hastened into the

house, closing and barring the door. The stranger, reaching the door, scratched and fumbled about for a considerable time … then slowly retreated towards the hills.

Mrs. Brennan confirmed the boys' statements regarding the man's appearance, stating that the stranger had long black hair, A search party afterwards scoured the hills, but saw no trace of the unwelcome visitor.

Case 128. July 1909. The Gulf, 20 km north of Emmaville, NSW

> A Bunyip, or What – A correspondent to the Tenterfield Courier states that again the strange animal has been seen by Mr. William Clifford, about five miles southeast of the Gulf. At first sight it was walking along and paused for a few seconds, then made off for the rough scrub country, and travelled at a great pace.
>
> Mr. Clifford's report … is the same as that of Mr. Wm. Carter, who saw the animal three times in succession in 1902, and Mr. Edward Clifford who fired three shots at the animal in 1906. (Case 120)
>
> The animal runs upright and stands about five or six feet high, thickly built and covered with long black hair, and its head is shaped like that of a dog, and appears to be very timid.

Richmond River Herald and Northern Districts Advertiser, 2 July 1909.

Case 129. 1911-1915. Pilliga Scrub, between Coonabarabran and Narrabri, NSW

In an article devoted mainly to the mysterious big cats that have been reported in various parts of Australia, a journalist, writing under the name of "Warrigal", made passing mention of gorilla-like creatures:

> The Illaroo gorilla was first reported in the Pilliga Scrub in 1911, but has not been seen since 1915.

"Warrigal," "Notes from a Countryman's Diary", *The Land*, 14 July 1933.

The Pilliga Scrub is still notorious for yowie activity. For a particularly dramatic modern-era report, see *The Yowie*, pp. 244-45.

Case 130. 12 October 1912. Between Bombala and Bemboka, NSW

George Summerell

On a bush track between Bombala and Bemboka, George Summerell rode up close to a strange animal that was crouched down, drinking from a creek. As it was covered in grey hair, his first thought was, "What an immense kangaroo", but when it stood up, he saw that it was about seven feet tall, with a face like that of an ape or man, minus forehead and chin. Its torso was all one size from shoulders to hips, and its arms reached almost to its ankles.

Seemingly unperturbed, it quietly returned Summerell's stare, stooped down again, finished its drink, then picked up a stick and walked steadily away up a slope.

One of Summerell's neighbours, the noted poet and bushman Sydney Wheeler Jephcott, rode to the site the following day and discovered about twenty footprints and handprints. Although the footprints "resembled an enormously long and ugly human foot" they showed imprints of only four toes. The handprints, too, were odd. They "differed from a large human hand chiefly in having the little fingers set much like the thumbs."

Two days later, he made plaster casts of one handprint and two footprints. Although he realised that the casts, made three days after the event, were less than perfect, he believed "that any reasonable being will be satisfied by an inspection … that something quite unknown [to] science remains to be brought to light."

Sydney Wheeler Jephcott

Unfortunately, however, Professor T. Edgeworth David of Sydney University, to whom Jephcott sent them, quickly decided (although he was a geologist rather than a zoologist) that the casts weren't going to set the scientific world on fire.

While examining them in the presence of an *Evening News* reporter, he said the imprints appeared to be "very human in shape … as though a man's foot had been put down, and then a hand added, the ball of the hand just obliterating the toes … the impressions are really very human."

Hastening to add that he had no doubt of Jephcott's sincerity, he pointed out that "the gentleman who sent the casts to me does not profess to have seen the animal."

Sydney Morning Herald, 16, 21, 23 and 24 Oct 1912. Also, *The Argus*, 28 Oct 1912 and many other papers.

Sir T. Edgeworth David

Regardless of whether or not Jephcott was mistaken about the tracks, there is no doubt that hairy ape-men, as described by George Summerell, have been reported in the very same area by many other people.

After reading of Summerell's experience, James Allen and R.W. Dawson came forward to describe sightings of similar creatures near of Creewah in 1875 and 1907. (Cases 43 and 124)

Yowies have been seen in the area much more recently. In April 1997, a ranger, Chris McKechnie, almost collided with a seven-foot-tall specimen only about four kilometres from the site of Summerell's encounter. (*The Yowie*, pp. 66-67). And in 1993 Mark Dowton, of Cooma, had a very close encounter with a slightly smaller creature at virtually the same spot. (Case 218)

Case 131. November 1912. Snowy River, NSW

"THE ALLEGED HAIRY MAN

> The mysterious hairy creature which some believe to be a man, reported to have been seen in the Bombala district a few weeks ago, has put in an appearance, it is alleged, on the Snowy River, on the upper reaches.
>
> Some men who have been fishing on the Snowy River, state that they saw a creature like a gorilla watching them intently. It carried a big stick, shaped like a nulla-nulla. After a time, it walked off into the bush. The anglers followed but owing to the mountainous nature of the country lost sight of the quarry.

Bairnsdale Advertiser and Tambo and Omeo Chronicle, 22 Nov 1912.

In the *Queanbeyan Age's* report of the same incident, on the same day, the word "gorilla" doesn't appear. Instead, the men are said to have encountered "a wild man very scantily attired". So, given that he was carrying a club (not common behaviour for yowies) perhaps this particular critter was actually just a human being gone feral.

Case 132. c.1913. Grafton district, NSW

In 1978, 81-year-old Annie Schenk described how her brother had seen a yowie in about 1913:

> ... my 18-year-old brother and my mother's cousin, a married man in his forties, went on a hunting trip in the Grafton Ranges. That's when they saw this creature – and saw him good and proper, too. It was no illusion because it was the frightened horses that put them on to the yowie's presence.
>
> They camped ... near a large pond, hobbled their horses and slept in a wagon, but at 4am the horses took fright and cleared out. At daylight they took out after them [and] rode them back to the camp site.
>
> ... as they got close ... the horses became frightened again and acted up. So the boys got off and led them in warily ... As they got near the lake, they saw this hairy, ape-like man pulling bulbs out of the water.
>
> He was rubbing them on his thigh, pulling off the roots and eating them. Luckily the wind was coming away from him so he was unaware he was being watched. But just as he got wind of something, he jumped up and took off for the thick scrub at the bottom of the range.
>
> The creature was a big six foot, maybe to six and a half foot [and] so much like a man they did not try to shoot him.

Unknown Grafton newspaper (?) 27 Jan 1978.

As Mrs. Schenk observed, the fear exhibited by their horses indicates that the men weren't simply hallucinating. Many other reports involve animals, both domestic and wild, being scared out of their wits by the Hairy Horrors.

Case 133. June 1913. Kendall, 10 km west of Laurieton, NSW

> Up at Kendall a number of residents have seen a wild gorilla or ape in the bush.

Durham and Gloucester Advertiser, 27 Jun 1913.

The village of Kendall is eight kilometres north of Middle Brother Mountain, where Eric Bond, his brother Len and others, encountered a huge yowie in 1973. (Case 194)

Case 134. September 1913. Daintree River, QLD

"A visitor to Cairns is credited with having circulated there a tall cock-and-bull story regarding the alleged experiences of a couple of prospectors at the Daintree.

The story ... is that the two men were washing off a prospect in a creek ... when a huge gorilla was seen approaching them. One of the men shot at and wounded the gorilla, which disappeared into the dense scrub. Later on, two gorillas emerged from the scrub and made for the men, but they shot and killed one, and the other made off again.

They skinned the dead gorilla and hung the skin up, and the dingoes got at It, but they did not damage it to any great extent. The visitors says that the skin is now at a selector's place at the Daintree, and further states that this particular place has always been avoided by the blacks as inhabited by "debbil-debbils."

A local scientist, who swallowed the tale whole, wished others to club together and send an urgent wire to the Premier urging him to immediately issue a proclamation prohibiting anyone from shooting at or otherwise Injuring any gorillas remaining "in their last vastness" ... [he] failed to find a sympathiser prepared to assist with the cash for the joint wire.

Daily Standard (Brisbane), 3 Sept 1913.

Although the above is almost certainly a tall story, it is worth noting that SAS Corporal J. Webster encountered a yowie at Dowrey Creek, a tributary of the Daintree, on 23 July 1985. (*The Yowie*, pp.154 and 296-97)

Case 135. November 1913. Near Eden, NSW

"Mr. Fred Alcock had an encounter with what he terms a hairy man at Mountain Top, Eden, one night lately. He was riding along when he noticed something coming towards him on all fours. On getting close to him it straightened up to its full height, and Fred's horse, aided by its rider, not choosing to wait and inspect, made a record trip to [the locality of] Greenland.

Northern Star (Lismore), 13 Nov 1913.

Yowies have been seen very close to the town of Eden in the modern era. (*The Yowie*, pp. 257-58 and 270-710)

Case 136. 1914. Near Suggan Buggan, VIC

In the 1950s, folklorist Aldo Massola learned about the Hairy Man from Victorian Aborigines.

> *Dulugars* were very strong, man-like, hairy beings. They lived in the mountains behind Suggan Buggan, and when women ventured alone in the bush they came flying through the air and took them away. The women were released after a while.
>
> Big Charley and his wife were walking along a bush track [just before the First World War] when they heard the three loud taps which always heralded a visit by the hairy men. Soon one of the creatures appeared, and Big Charley stood his ground and prepared to fight, while his wife ran into a cluster of gum trees for safety. The clustered trees were close together, and the Dulugar was unable to reach the woman. He was also worried by the husband, and eventually retired.
>
> Big Charley, bleeding from a profusion of wounds, and his wife, with her dress almost torn from her where the Dulugar had got a hold, were able to reach Lake Tyers without any further molestation.

Aldo Massola, *Bunjil's Cave: Myths, Legends and Superstitions of the Aborigines of South-East Australia*, Lansdowne Press, Melbourne 1968.

This story is interesting for two reasons. It is the only item of Aboriginal lore we know of that refers to yowies having the magical power of flight, and it also features the earliest reference we know of to a Hairy Man signalling his approach by "three loud taps." Some modern-era yowie researchers (and, indeed, several North American sasquatch hunters) claim to have heard such rapping – like a tree trunk being hit three times with a stick – in Hairy Man hot spots.

Case 137. September 1918. Mangrove Creek, 12 km west of Gosford, NSW

> A correspondent writes to the 'Gosford Times':- I was making my way along Mangrove Creek the other day when I came across some strange looking tracks fully 18 inches long. The longer you looked the longer they seemed to grow. I followed them for nearly half-a-mile when all of a sudden they disappeared in a scrub and I wasn't game to follow them in there.
>
> That evening just about dusk I saw a frightful looking thing bounding along

the rocks on the side of the range. It looked to be covered in hair like a black bear of the grizzly sort, and next morning I went and found the tracks. If it ain't a 'hairy man' what the blanky blank is it?

Windsor and Richmond Gazette, 6 Sept 1918.

This story is of special interest because Mangrove Creek is where two youngsters, a brother and sister, encountered a dark, hairy, eight-foot-tall "yahoo" in April 1843. (Case 14)

Case 138. September 1919. Nulla Nulla, 50 km north-west of Kempsey, NSW

"Some consternation has been caused amongst the residents, of Nulla of late, by the presence of a strange animal, described by some as an ape without a tail, by some as a gorilla, and by others as a bear with a long bushy tail. The animal moves about on all fours, but stands up on hind legs at the approach of anyone. It is said that its presence is known by the odour, before being seen.

To capture this ferocious animal a party was organised … After searching Mount Anderson and the vicinity for several hours, the hunters returned home after a fruitless attempt.

Gloucester Advocate, 27 Sept 1919.

Four years later, a timber-feller, Jack Brewer, encountered another (or the same) Hairy Man at Nulla Nulla Creek. He was so traumatised that he vowed to never again enter the forest. (Case 142)

Case 139. November 1919. Creewah, 20 km north of Bombala, NSW

"THE HAIRY MAN – ONCE MORE

A Tantawanglo resident, who has a grass run near Creewah, states positively that he, with others, saw an animal in a swamp with some cattle. It appeared to be feeding, and was on all fours until disturbed, when it stood upright like a human being. It was covered with long reddish hair, and had a big head and very large mouth. It turned over with ease a large log as if looking for grubs to eat.

When it sighted the visitors it made toward them, but they were not looking for a closer acquaintance, and, jumping on their horses, made off.

A search party is to be organized at Candelo to try and capture the (as yet) mythical creature. In captivity it should be worth its weight in gold. An enterprising [resident of Bega] has offered £1000 for the animal if taken alive.

The Bega Budget, 19 Nov 1919.

As we have seen, similar hairy horrors were seen in the vicinity of Creewah in 1895, 1907 and 1912. (Cases 82, 124 and 130)

Case 140. 1920-1925. Cox's River, Blue Mountains, NSW

In the early 1920s several people – including the well-known businessman Mark Foy – reported encountering a huge wild man or gorilla in the vicinity of Cox's River and Black Jerrys Ridge, in the Megalong Valley.

"WILD MAN SEEN

Last weekend a party of shooters set forth … to seek foxes and rabbits on the banks of Cox's River. The members … were Messrs. Mark Foy, A. J. Birney, Hugh Mocken and Dr. Pitter.

[They] had reached the foot of Black Jerry, a well-known mountain in the Valley of Megalong, in the vicinity of the river, when suddenly a form leapt from a tree, and sped swiftly into the … neighbouring scrub. It was absolutely nude, and hair a yard in length streamed from its head. The visitors scarcely had time to ascertain that the apparition was a well-formed man, about six feet in height, with a hairy torso, before it had vanished into the shadows of the undergrowth.

Investigations revealed the tracks of a huge splay foot, with toes set wide apart, as is the manner of men unaccustomed to boots. There were several imprints of the bare foot in sandy patches, and the party are agreed that the foot was abnormally large, even for a six-foot man of heavy build. … the evidence of four reputable eye-witnesses cannot be dismissed lightly. – Katoomba 'Echo.'

The Blue Mountains Echo, 27 Feb 1925.

A follow-up appeared two weeks later:

Mark Foy

"BLACK JERRY

A party is to be formed ... which will proceed to the locality of Black Jerry, and endeavour to elucidate the mystery of the "hairy man."

Mr. Cowling, who lives near the haunts of the wild man, places little credence in the story. He has roamed in the district for years, and has never seen trace of the stranger.

Mr. Cooper is not only a believer in the story, he swears he has seen the "thing' twice within the last five years and on both occasions near the foot of Black Jerry. On the first occasion, the horse he was riding shied violently and he was unseated. Two years later, he was walking through the bush, when he distinctly saw the hairy one leisurely walk from a rock into the scrub. Cooper said he was far too big to argue with. He does not think it is a man: he thinks it is a great gorilla of apparently wonderful strength.

Mr. Noel Patterson, of Katoomba, swears he saw the big thing about 18 months ago. He reported it to Jack Scrimgeour and others at the time, they treated it as a joke, putting it down to a fevered imagination.

Adelong Argus, Tumut and Gundagai Advertiser, 12 Mar 1925.

Several yowie sightings have occurred in the same area in the modern era. (*The Yowie*, pp. 165-67.)

Case 141. 1923. Watsons Creek, 42 km north of Tamworth, NSW

While working in rugged bushland, Tamworth grazier Henry O'Dell spotted large footprints in the sand and then spied a strange animal hanging by one arm from a tree 30 yards away. It seemed about six feet tall. Although it was extremely hairy, its eyes, mouth and nose were visible. His workmate, Keith Blanch, also saw the creature before both quickly retreated.

O'Dell also said that a friend, Tom Chapman, had once shot a female "ape-creature" at Wild Cattle Creek before fleeing in fear. He left the carcase where it lay.

Psychic Australian, August 1977. Vol 2, No 8.

The *Psychic Australian* journalist mentioned that the story originally appeared in a Tamworth paper, but we haven't managed, so far, to locate the original article.

Quite apart from the eyebrow-raising assertion that an "ape-creature" was killed there and left to rot, it is interesting to note that Watsons Creek, like a considerable number of other supposed yowie locations, was also once noted for out-of-place big cat sightings. In 1914 the district was in a state of alarm over repeated sightings of a "lion".

Case 142. 1923. Nulla Nulla Creek, about 50 km north-west of Kempsey, NSW

Retired timber-feller Sam Chapman of Port Macquarie told Patricia Riggs, of the *Macleay Argus*, that in 1923 a workmate, Jack Brewer, walked into their bush camp, kicked out the fire and announced he would never go into the surrounding forest again. He had seen a Hairy Man walking along a log.

"He never did go back," said Mr Chapman, "I had to get some contractors to come in and finish the job."

Macleay Argus, 28 Sept 1976.

Four years earlier, in September 1919, armed riders searched the same area after sightings of a huge, smelly "ape", "gorilla" or "bear". (Case 138)

Case 143. c. October 1923. Einasleigh, 220 km west of Ingham, QLD

> "PRE-HISTORIC" (?) ANIMAL AT EINASLEIGH
>
> An Einasleigh correspondent writes: Quite a sensation was caused here recently when a pumpman ... saw at the falls a strange animal, which he believed to be a bunyip. The pool is at the bottom of the falls and is known as Cleopatra's Bath.
>
> ... the animal resembled a bear with heavy limbs and a thick shaggy mane down the back. It had no tail, but a savage head resembling a bull calf. As he approached [it] retreated into the caves under the cliffs.
>
> Since then it has been frequently seen by the local J.P. and the sanitary contractor and others, only by night, floundering in the bath ... some of the railway

> men declared they would capture it ... But when they saw him and heard him grunt they funked the job. They have come to the conclusion that the stranger is some pre-historic animal.

Cairns Post, 5 Nov 1923.

Although the description is very sketchy, this bear-like "bunyip" *might* have been a yowie. In any case, the incident is interesting because the river in which the creature wallowed, the Einasleigh, also featured in the "wood devil" incident of the 1880s, during which Arthur Bicknell claimed to have shot and killed an aggressive, man-sized monkey. (Case 80)

Case 144. c. 1924. Southern Highlands, NSW

In 1987, Val Whalan of Huskisson told how her grandfather and four of his children, including Val's mother, encountered a yowie many years earlier, in a remote valley on the Southern Highlands.

It happened one night as they were checking their rabbit traps. After hearing a crashing sound in the bush, they hid and "smelled a terrible, foul odour, and saw a huge, hairy beast like an ape; his smell was blowing to them so he didn't know they were there. He came across the creek within yards of them. They were terrified, not moving until the beast was well away.

"My Granddad would not go around the traps at night after that."

Telegraph (Sydney) undated clipping, probably 7 Aug 1987.

Case 145. 1925. Near Murray Bridge, SA

Mrs P. Lindsay said that when she was 11 years old, she saw a creature known to Aborigines of the lower Murray as *mooluwonk*. It was more than 10 feet tall, had long black hair, dark red eyes, large teeth and, as it spent much of its time in the river, webbed feet and hands.

Under the direction of Mrs Lindsay and seven other Aboriginal elders, artist Brian Vercoe produced an excellent sketch of the creature.

The Advertiser (Adelaide), 5 Jul, 1973.

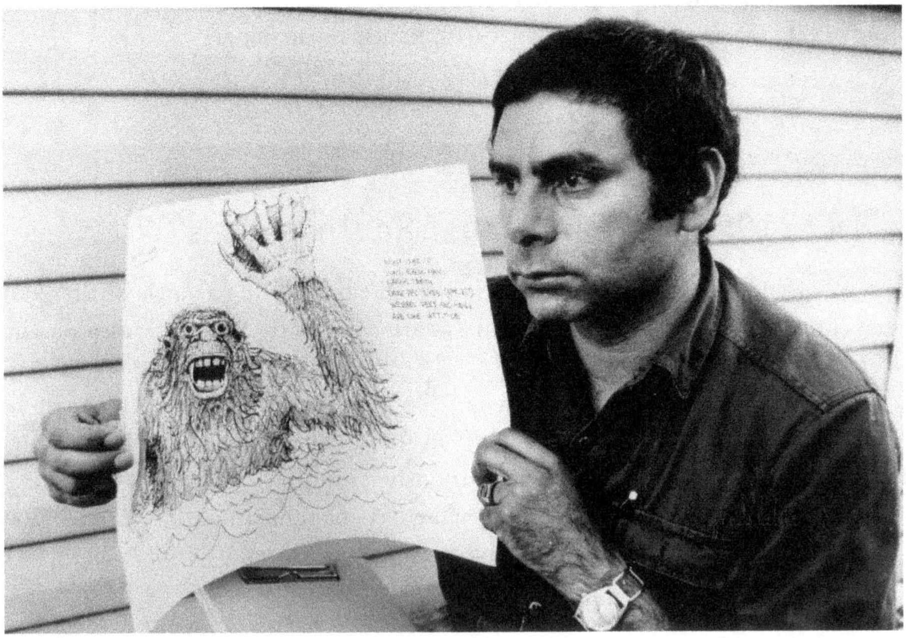

Brian Vercoe and his sketch of the mooluwonk.

Case 146. 1928. Near Palen Creek, south-east QLD

In 1980 Bob Mitchell, then 76, of Redcliffe, told of seeing two yowies at close quarters 52 years earlier.

The first sighting occurred as he and two mates were trekking on horseback through thick bush, just north of the McPherson Range.

> It was about 10 am – the yowie was standing in a clearing not far from us and in that light there was no mistaking it for anything else. It was about seven feet tall, with a black human face and a gorilla-like body covered in thick brownish hair.
>
> It showed no aggression; it just looked at us for a moment, then turned and disappeared into the bush. It had really big feet and could move fast.

A few weeks later the men were camped near Widgee Mountain, about 32 kilometres to the east.

> We saw another yowie – it too just looked at us for a moment, then disappeared.

I'm speaking up now because I want to protect the yowie ... The destruction of wildlife in this country by trigger-happy men is appalling."

Sunday Mail (Brisbane), 9 Nov 1980.

Case 147. 1920s or early '30s. South-east NSW

Fred Howell often told his grandson, Billy Southwell, about seeing a yowie on an old bullock track that climbed from the New South Wales south coast to the top of the escarpment to the east of Braidwood.

On reaching a fork in the road known as "The Three Ways", Fred and his mate hobbled their team and took buckets to a waterhole. On the way back their blue cattle dog charged ahead and began circling the dray, barking furiously. On top was a very angry looking Hairy Man – Fred referred to it as a "yourie" – throwing bags of grain around. It jumped down and ran into the scrub. Fred saw what might have been the same creature at the same spot on one or two other occasions.

Tony Healy, interviews with Billy Southwell, 1986 and 1998.

Case 148. c.1930s. Burnt Bridge, near Kempsey, NSW

When she was ten years old, Mrs Mamie Mason (then Mamie Moseley) encountered a Hairy Man quite close to where Chris Davis (Case 109) was abducted some years earlier:

> "My two young cousins, Tom Campbell and Zelma Moseley, were with me … we were coming home … just on dark. I had a little dog with me. We were going past the lantana and I could hear this growling … and sticks cracking. That little dog was yelping and going on … its hair was standing on end.
>
> I pushed [the little ones] under the fence just as it came through the lantana. They rolled down the embankment … and I stood there and this thing … it had hair all over it and the smell was something terrible … something like a pig, but not quite … a sour, stale stench, like something rotten. It came towards me ambling, moving its arms.
>
> It had long arms. About down to its knees. Long hands … and it stood looking at me. I could see it had a face … not like an animal, and [hair] down over its eyes, and the eyes were glaring. It sort of had me hypnotised. My little dog butted me on the legs … I jumped the fence, rolled down … and ran. The other

two kiddies … started screaming. My grandfather, John Moseley … ran down and shot the gun off. I couldn't talk … the little dog crawled under the house … wouldn't come out until the next morning.

Macleay Argus (Kempsey), 2 Oct 1976. Credit: Patricia Riggs.

Case 149. 1930. Maria River, about 10 km south of Kempsey, NSW

When she was 12 years old, Melba Cullen encountered a yowie near her home on the Maria River.

She was interviewed by journalist Patricia Riggs in September 1976 and by the authors in 2001. On both occasions, she stressed that the experience was still vividly engraved on her mind. So clear was the memory, that she was confident the sketch she did for us, 71 years after the event, was a reasonable likeness of the creature.

Melba Cullen's sketch

> I loved collecting flowers and … walked out of the bush into a clear patch … I heard heavy footsteps a few paces behind, but I thought it was my brother … I looked back … there was a big stump just through the fence near me … suddenly I saw this huge Hairy Man looking around that big stump.
>
> He was about seven feet tall and very broad shouldered. He had long, tan-coloured hair all over him. The hair on his face was about as thick as the hair on a dog … I took one look at him and ran away screaming. I swear to this day the thing … was a real Hairy Man. It wasn't a kangaroo and wasn't an ape or a monkey. It stood up straight like a man.

Macleay Argus, 25 Sept 1976. Witness interview with Paul Cropper, 15 Jul 2001.

Case 150. April 1930. Dunmore, 2 km south of Shell Harbour, NSW

"Says the Kiama 'Independent':- A considerable scare is felt on some farms at Dunmore in the vicinity of Connolly's range, as some strange animal prowls in the vicinity, and has at a distance been seen above Terrangong Swamp. While fox shooting, Mr. N. Hambly got the best glimpse of it and thinks it is a wild bear, that perhaps has escaped from some menagerie or circus.

Moss Vale Post, 25 Apr 1930.

Because the region has produced many reports of hairy ape-men, it seems likely that those who glimpsed the Dunmore "bear" actually saw a yowie.

Case 151. June 1930. Nullica River, 7 km south-west of Eden, NSW

"STRANGE ANIMAL AT NULLICA. WHAT IS IT?

Children of the Nullica River settlement have been scared and excited by the appearance in proximity to their homes of a strange animal, brown in colour, much larger than the average cattle dog and resembling a monkey in shape. That, at least, is the description given of it by Tommy Bobbin, the biggest of the two boys who have seen it. He has seen it on several occasions, once at a distance of only a few feet.

The first time he saw it the animal was sitting up in a gorilla-like attitude, with what appeared to be a stick in one of its hands. Terrified, but brave, Tommy threw at it a cob of corn ... thereupon the horrible looking creature disappeared into the scrub. Subsequently it was seen by other children, and they ran homeward, screaming hysterically. All attempts to convince [them] that the animal may have been a strange dog are scouted by the children, some of whom at least are old enough to know the difference between a monkey and a dog. The parents have so far not sighted the animal, the identity of which is so far a complete mystery. – Eden Magnet.

Delegate Argus, 19 Jun 1930 (quoting the *Eden Magnet*).

Case 152. 1930s. Epping, NSW

A local man said that, as a boy, he'd seen a bear-like creature sitting on a rock in the bush at Epping. Now in Sydney's inner western suburbs, Epping was then near the extreme western edge of the city.

The animal was five to six feet tall and definitely out of the ordinary.

Witness interview with Paul Cropper, 1982.

Case 153. 1930s. Megalong Valley, NSW

Mrs. Lola Irish wrote that while she and her brother were holidaying in Katoomba in the 1930s, their landlady told them about her sighting of a "giant hairy ape-man."

While returning to a campsite near the "Ruined Castle" rock formation in Megalong Valley, she'd seen the creature disappearing into the bush, carrying off some stores from the campsite.

Letter from Lola Irish to the *Sydney Morning Herald*, 14 Sept 1978.

Case 154. 1931. North of Moore River, WA

In August 1931 three young Aboriginal girls, Molly, Gracie and Daisy, escaped from the Moore River Native Settlement and in an epic journey walked 1600 kilometres north to Jigalong, on the edge of the Little Sandy Desert

One afternoon, as a storm was brewing, they were startled by the sound of heavy footfalls coming their way. As they lay hidden in a thicket, the footsteps came so close that they could feel the ground vibrating. Suddenly a huge man-like creature emerged from the banksia scrub and ran past. He was of seemingly gigantic stature and his massive footfalls were audible well after he disappeared from sight.

"He was a big one alright," said Daisy later, "... had a funny head and long hair."

"It was a marbu alright," agreed Molly. "A proper marbu."

Doris Pilkington/Nugi Garimara, *Rabbit Proof Fence*, pp. 84-86.

Case 155. c. 1932. Lake Condah, VIC

Aldo Massola

In the 1950s, Indigenous people at Lake Condah told folklorist Aldo Massola about little men they called *net-nets*.

The *net-nets* (presumably the creatures known elsewhere as junjudees) were very small, hairy, and had claws instead of finger and toenails. They were mischievous but harmless, and were believed to live in natural hollows among jumbled heaps of boulders.

Andrew Arden told of encountering one in about 1932, while hunting with his wife in the Stony Rises near the lake. He had just shot a rabbit when "one of the little people" suddenly appeared, seized the carcass, and ran away over the rocks. Mr Arden gave chase but soon lost sight of the light-fingered Lilliputian.

Aldo Massola, *Bunjil's Cave: Myths, Legends and Superstitions of the Aborigines of South-East Australia,* p.150.

Case 156. c. 1932. Gundiah, 35 km south of Maryborough, QLD

Just as the yowie phenomenon has occasionally been confused with the bunyip (water monster) legend, it has also sometimes been confabulated with another zoological mystery – the riddle of seemingly uncatchable big cats – "lions," "cougars" and (most often) "black panthers" – that have been reported for 150 years or so in the Australian bush.

> "A SUGGESTION that more than one strange animal is at large in the Wide Bay district has been made by a correspondent, Mr. James Cavanagh, of Gundiah.
>
> Mr. Cavanagh states: "I hope my letter may assist in hunting down this strange animal known as the Yengarie "Lion." I wish to point out that it is very foolish for people to run away with the idea that the animal is a large dingo.
>
> "From my own experience, I am satisfied that there is more than one animal prowling around the Wide Bay District. Some 14 years back I complained to the Tiaro Police of two strange animals in my paddock, which to me seemed to be large monkeys or gorillas.
>
> "These animals, or some animal, gave me considerable trouble as regards the killing of my pigs ... What pigs were not taken away had the lives crushed out of them ... This animal might pay a visit once a month, or ... every night for a week. ... I have been forced to sleep on the roof of my pig stye ... to try and locate the animal, but he is too cunning. Whatever [it] is, there is no dog that will face it.
>
> I definitely say that it is not a dingo, as I have had twelve-month-old heifers killed ... Every dingo is a head and flank killer, but this animal killed to get at the jugular vein and drank the blood and never touched the body."

Maryborough Chronicle, 16 July 1946.

Mr. Cavanagh isn't alone in reporting "gorillas" near Gundiah. A huge (ten or eleven-foot tall) yowie scared the daylights out of Stan Pappin and his mate Billy Wilson, just three kilometres to the north-east in 1987. (*The Yowie*, pp. 260-61)

Case 157. January-February 1932. Between Bright and Yackandandah, VIC

Late one January night, as he rode home to his property at Running Creek, a farmer, Mr. Cherry, received a terrible fright. On 27 February, *The Mail* (Adelaide) ran the story under an eye-catching headline:

> "ARMED MEN HUNT FOR STRANGE MONSTER
> SHAGGY BEAST LIKE GORILLA ATTACKS THREE
>
> Armed men, the article (and several follow-up stories) revealed, were "hunting for a mysterious beast, shaggy and powerful … in the mountainous region between Bright and Yackandandah."
>
> The drama began late one night about a month previously, when Mr Cherry, a farmer, of Running Creek, was riding home:
>
> [He] came to a gate, bent from the saddle to lift the catch … there was a grunt and a scuffle and a bulky beast leapt at the horse's head. The horse bolted with Cherry clinging to its back. Next morning … he discerned prints that suggested the foot of a grizzly bear.
>
> A few nights later [he] heard the strange visitor lumbering and grunting around his hut. Outside, the horses whinnied in terror.
>
> Three of the most daring men … spent [a] night out … With guns ready. [On] Running Creek road … the thing they had been seeking was on them. The horses … rearing, broke the shafts of the buckboard … the men were thrown to the roadway … they fired, but … the animal scuttled back into the bush, apparently unharmed.
>
> On Thursday [25 February] William Nuttall, a 21-year-old drover, with some women and men friends, was riding home to Myrtleford. The moon was shining. [Near Eurobin railway station] young Nuttall got off his horse to tighten the girth, and the others went on slowly.
>
> Those ahead heard him shout, "Ride like mad! Some strange beast is attacking me." [It] had made a sweep at him with its paw, [ripped his] shirt to ribbons, but missed his body. [His] horse … bolted, but he stuck to the saddle.
>
> When [he] looked around he saw a large hairy creature. A sudden swerve of the horse unseated him, and … He raced for life … with the animal close on his heels. About 20 or 30 yards ahead he saw his horse standing on the roadway, shivering with fright. [He] regained the saddle, urged the horse on.

An article in the *Argus* (Melbourne) included a description of the beast, supposedly supplied by Mr. Nuttall:

> [He] described the animal as being about 7ft. in height, with [a] round, hairy head and four tusks. It stood on two legs. It is believed to [have] escaped from a travelling circus when it was at Yackandandah some time ago. The animal has been seen by residents in different parts of the district, its tracks being plainly discernible. It is said to resemble an ape.

The Mail (Adelaide), 27 Feb 1932; *Argus*, 27 Feb 1932;
Sunday Times (Perth), 28 Feb 1932; *Evening Post*, 15 Mar 1932.

Apart from its bizarre dentition, this unusually feisty animal seems to have been reasonably yowie-like. Hairy ape-man reports have emanated from the same general area in more recent times. (*The Yowie*, pp. 267 and 296.)

Case 158. 1932. Wollondilly River, south-east NSW

While on a shooting trip, Chris Bagnall and some friends camped next to the Wollondilly River.

One bright, moonlit night, while washing dishes in the stream, Chris looked up to see a tall, hairy man standing on a boulder on the opposite bank, watching him. He dashed back to the camp to grab his rifle, but on his return the creature had vanished.

Many years later, he related the incident to his grandson, Mick Stubbs, who passed it on to researcher Gary Opit.

Credit: Mick Stubbs and Gary Opit.

Case 159. February 1933. Darnum, 5 km east of Warragul, VIC

Coincidentally, this case, and the one immediately following it, are the first reports we know of in which yowies (or, at least, yowie-like creatures) were sighted from motor vehicles.

> A STRANGE ANIMAL
> Residents Alarmed

A strange animal, supposed to be a species of bear, has caused residents of Darnum district some concern. Residents have seen tracks, unlike those of any native animal, resembling a huge hand or claw. One man has heard weird cries at night, and his horse … became frantic.

On Sunday night a carrier from Warragul caught sight of a huge black creature shuffling across a bush track, but it was soon beyond the range of his headlights and disappeared in the scrub.

Townsville Daily Bulletin, 18 Feb 1933.

Case 160. July 1933. Between Tinonee and Krambach, NSW

"THAT STRANGE ANIMAL

Word comes to hand that a strange animal was seen along the road between Tinonee and Krambach, about a mile on the other side of the Hillview turn-off.

Mr. Richard W. R. Hall, accountant, of … Sydney, was coming to Taree in his car. At about 7.30 on Wednesday night, at the point stated, he saw an object on the road. He could only see the head, forequarters and front legs by the aid of his headlights. His first impressions were that it was an animal of the Polar bear species, but as the car approached, the animal, in two jumps, disappeared in the darkness of the bush.

Mr. Hall told our representative that he has been travelling in cars, on all sorts of bush roads, for the past ten years and knows by sight any animal common to the bush. This beast was none of them, and, whatever it was, he is confident that it was "no animal which belongs to Australia." – M.R. Times.

Townsville Daily Bulletin, 18 Feb 1933.

In the modern era, the Krambach-Taree area became something of a Hairy Man hot-spot. One 1990 episode, during which 13-year-old Julie Clark and her friend Jodie Betts twice encountered hairy, hulking yowies while horse riding near Krambach, was particularly interesting. (*The Yowie*, p. 265)

Case 161. August 1933. Denman, 26 km south-west of Muswellbrook, NSW

Although this story concerns an admitted hoax, we think it worth including by way of comparison with our many other more genuine (we hope) "gorilla"/yowie reports.

The village of Denman is located beside the Hunter River, a stream that was mentioned, in connection with the Hairy Man, as far back as the 1830s.

> JUST AN IMITATION! DENMAN'S "GORILLA"
>
> Several reports of a strange animal, resembling a gorilla, having been seen round the township of late, were received by residents with mixed feelings. Investigation proved that there was some substance in the reports, but the gorilla was a member of the Jungle ball committee working an advertising stunt for a ball to be held shortly.

Muswellbrook Chronicle, 25 Aug 1933.

Case 162. February or March 1934. Near Stanthorpe, QLD

Cecil Thompson was born and raised on a farm off Sugarloaf Road, just out of Stanthorpe, "near the two-mile peg."

One day, when he was 12 years old, he and his brother Ernie worked for hours carrying bags of peas up to the homestead from the river flat. They left one bag near the river, and after tea their father told them to go back down and fetch it.

> It was in the dusk. We saw a big form bending over the bag, and it straightened up. I thought it was George Wells or someone dressed up in a suit having a joke, so I yelled out, "Whoever you are, you don't scare me!" It took no notice at first, so I threw a clod of earth and it took that sort of forward and sideways movement, with that sort of guttural "Woonk, woonk, woonk." And it had taken off by this time across a drain ... and I took off too, and passed Ern, and we got back to Grandfather's place out of breath.
>
> It was ... between five foot six and six feet, covered with dangling hair, but not on the face. It looked like a person dressed up in a suit, but of course it wasn't. It had dangling arms, and the forward and sideways movement was a bit

like Charlie Chaplin used to walk – sort of a waddle, but not exactly, because there was a sort of stamp or stomp at the same time.

The Stanthorpe Border Post, 1 Nov (probably 1992); Cecil Thompson interview with Paul Cropper, 1997.

Over the years Cecil interviewed several friends and relations who encountered similar creatures in the Stanthorpe-Eukey area. (Cases 163, 172 and 174)

Case 163. c. 1935. Near Stanthorpe, QLD.

In 1997, Cecil Thomson recalled his cousin Walter Beddoe's encounter with a yowie:

> [He] was coming to town with his horse one afternoon. It was a bit misty and he struck a Hairy Man on the Stanthorpe to Mt Tully road, near what we now call Hairy Man's Bend. It was only about three feet away and made a grab at the bridle and then strode over the fence. It was apparently tall enough to stride over the top wire.

Cecil Thompson interview with Paul Cropper, 1997.

Case 164. 1935. South-east NSW. Aboriginal lore

> "DOULIGAH."
> (By "Eureka")
>
> Although there are few full-blooded aborigines remaining on the south coast of New South Wales, you can still hear all about the "Douligah", or the hairy man … Every aboriginal on the south coast firmly believed in the evidence of the douligah, and they dreaded him … They describe [him] as a man of powerful build, capable of tearing down small trees and lifting great rocks … hair all over his body, and though he remained in the mountains during daylight, he frequently visited the abos' camps at night, [terrifying them].
>
> The natives about Nowra had seen one while those at Twofold Bay believed one lived in the mountains there. We are apt to discredit these stories … but it is quite possible that such a creature did exist in the early days. Perhaps it was a kind of gorilla.

Sydney Morning Herald, 8 Jun 1935.

Case 165. 1935. Lismore, NSW

In July 1977, a 52-year-old man told journalist Gary Buchanan of seeing a yowie on his grandfather's farm in South Lismore 42 years earlier:

> "I was standing on the veranda ... when I saw a man walking across the paddock from the direction of the hills ... as it walked towards the house, my grandfather's horse started to kick up a hell of a fuss.
>
> When [his grandfather] saw what it was he pushed me inside, blew out the lamp [and] grabbed his rifle ... we all watched through a small window as the creature walked past ... [about 25 yards away and clearly visible in the moonlight] ... its head didn't seem to have a neck, but was sitting straight on its shoulders. It also looked as if it had a hunched back, but it was standing up straight. It was much thicker around the shoulders and chest than a man ... my grandfather told me it was the same creature he had seen ... only a few years earlier. He had ridden into [a] gully to pick some guavas when he saw [it] come down one side of the gully, cross a small creek, then climb up the other side of the hill.

Lismore Northern Star, 7 Jul 1977.

Case 166. 1937. Nana Glen, north-east NSW. Aboriginal lore

The following unsigned article (here heavily edited) appeared in the June 1937 edition of *Mankind*.

> "THE YERRI-WAHOO
>
> J.N., a Kumbangerai, told me that big hairy men lived in the scrub at Nana Glen ... and were called ... Jarra-wahu.
>
> On one occasion a kuri (aboriginal man) who found a turkey's nest made a fire to cook the eggs ... a Jarra-wahu arrived [they fought over the eggs and the man hit the creature's leg with a tomahawk].
>
> As he was struck, he cried out "Koin", which brought up his companions. They charged the [Aboriginal man] and to impede his retreat, threw a fog in front of him, but he escaped.

> The Jarra-wahu killed [Aborigines] if they caught them.

Mankind Magazine, Jun 1937.

Nana Glen has been the site of several recent yowie incidents. (e.g., Case 216)

Case 167. 1930s and '40s. Deua River, NSW

A steep and wild region to the west of Moruya is drained by the serpentine Deua River. In the 1930s and '40s, Rodney Knowles, whose family owned land beside the lower reaches of the stream, about 15 miles inland from Moruya, often heard old timers speak of something they called the yahoo, yowroo or yowrie.

Twice within about 20 years, he found footprints that "could not be identified as being those of any known animal." They were five-toed and "of considerable depth in river sand … the distance between them was up to seven feet [213 cm]."

He found both sets of footprints in a large gully that comes down to join the Deua from the west. For as long as he could remember it had been known as Yahoo Valley.

Queanbeyan Age, Sept 1976 (exact date unknown). Witness interview with Tony Healy, Sept 1976.

About 35 kilometres south of the Deua is another, much smaller stream – the Yowrie [sometimes spelt Yourie] River. The watercourse was known by that name as early as 1885 and hairy ape-men continue to be encountered in the vicinity. (e.g., Case 225)

Case 168. c. 1930s or '40s. Dungay Creek, 25 km west of Kempsey, NSW

When she was about 10 or 11 years old, Mrs Joan Delaforce (nee Clarke) was walking with her sister and brother near the family farm when one of their dogs began acting strangely.

> [It] … ran up into the bush and started barking and growling as if it was scared. When we got a bit closer, we got a good look at this thing. It was standing about six feet tall and was dark brown in colour with long hair about the neck. We didn't get a good look at the face. We didn't stay around long enough.

On another occasion, Joan and her siblings saw something moving about in the tractor shed. As they approached, the creature they'd seen earlier came out and walked away.

Another incident occurred as they were walking to the school bus stop:

> I suppose the distance was about half a mile ... on the next hill. What we saw were two big creatures and three small ones. That was the last time I saw them, but my brother and sister have seen them since. When you have seen it, you will never forget about it ... I am telling you that there *is* a Hairy Man ... in the bush.

Macleay Argus, 2 Oct 1976. Credit: Patricia Riggs.

Case 169. c. 1930s or '40s. Sebastopol, 40 km north-west of Kempsey, NSW

Tom Carroll, of Gladstone, told the *Macleay Argus* that his late uncle, Joe Carroll, saw a Hairy Man while shooting in the scrub near Sebastopol:

> Joe was ... almost within 50ft when he saw this creature about 5ft tall with long hair and heavily built with long arms. When the creature saw him, he got up from a sitting position and walked away with a steady gait.

On the following day Joe took some men from Moparrabah to the site and they followed the creature's tracks for some distance. A degree of corroboration came from Leslie McMaugh, a local landowner, who said his horses sometimes snorted and galloped wildly around his paddock at night, as if frightened by something. Sebastopol is 16 kilometres north of Kookaburra, where George Gray was attacked by a small ape-man in 1968. (Case 190)

Macleay Argus, 14 Oct 1976. Credit: Patricia Riggs.

Case 170. 1937-41. Randallstown, Cockatoo Valley and Barossa goldfields, SA

> THE GORILLA FANTASY
> Cockatoo Valley now gets into the news
>
> Mr. John Zerk, of Cockatoo Valley, writes about the gorilla fantasy supposed to exist ... near Angaston:

"This same animal, or rather one which would compare with the description … has been in our Cockatoo Valley and Barossa goldfields' locality for quite three or four years. It has been seen by quite a number of local people, and is of a very timid nature, as it hardly gives anyone enough time to obtain a really good look at it.

"It appears [to have] escaped from the late Mr. Bellchambers' menagerie at Humbug Scrub years ago, and seems to be … fruit-eating … as it has been seen in a vineyard not 200 yards from my place. In fact, my sister saw it near our horse stable about 12 months ago.

"It is known in our district as 'the black animal' and does not appear to be of a destructive nature."

The Bunyip (Gawler), 20 June 1941. Also, *The Recorder* (Port Pirie), 27 Jun 1941 and *The Bunyip*, 23 and 30 May 1941.

Case 171. 1938 or '39. Between Nanango and Maidenwell, QLD

Clyde Shepherdson

When Clyde Shepherdson was 13 or 14 years old, he had an unforgettable experience:

> "There was a lot of heavy vine scrub up there in them days ... Me and a mate, Clarrie Parsons, had a couple of single-shot .22 pea rifles and ... we used to go shooting little red-backed scrub wallabies. You had to be pretty quiet or you wouldn't see them, so we were [moving quietly] and I say to this day that's the only reason we came across him.
>
> It was deep into the scrub, a couple of miles from the nearest road, and we were creeping through, bending down [under the vines] and we got to a little open space and stood up, and there, about 20 feet away, this great big yellow thing was standing there with its back up against a fair lump of a tree – and I think he got a bigger fright than we got!
>
> He put his hands up on either side of his head ... with his palms towards us and snarled, more or less to say, "You'd better get out of here!" His fingers were curled, like a claw – like it was going to grab you or rip you open. I suppose we could have shot him but we never even had it in our mind ... we just thought, like, "the quicker we get out of here the better!" We just wandered off slowly ... keeping our eye on it, didn't turn our backs. We didn't want to run because he might have took after us. It kept its back against the tree, followed us with its eyes until we got out of sight. Once we got out of the thick scrub we just bolted!
>
> It was a rusty yellow ... camel-colour would be the closest. I've still got him imprinted on my mind to this day. It would be a good six foot ... shaped more or less like a person only very broad across the shoulders; fairly nuggetty; solid. And it had a fair amount of fur on it – more like fur than hair. It was pretty rough looking but not too dishevelled, not too dirty. Fairly long arms ... its legs were furry all the way down; they were pretty wide.
>
> It seemed to have a fairly broad forehead ... it looked real ape-like – definitely an animal impression. The face was fairly flat and broad. You see those drawings of stone-age men: it looked a bit like one of them. It had a short neck. We saw its teeth when it growled ... they were more fangs than teeth ... fairly long – savage. It had fairly big ears, covered with hair but sticking out – you couldn't miss them. A flattish nose. The skin looked dark on the face but it had a lot of hair on it – scruffy-looking hair on its head.
>
> It was a pretty cloudy and damp day, but it was after lunch, in full daylight. He was out in the open – there's no two ways about that. I've lived with this for 60 years, and now I can finally tell someone about it. I'll never forget it – never!

Witness communication with Dean Harrison of AYR;
witness interviews with Paul Cropper and Tony Healy, Apr and Jul 2001.

Case 172. c. 1940. Near Eukey, QLD

Cecil Thomson (Cases 162 and 163) recalled how his older cousin, Teddy Collie, saw what was apparently a yowie:

> [He] came home sweating; his horse was sweating. As he put it, he'd seen a "big baboon" on the road from Ballandean to Eukey. The spot is still known as "Baboon Gully."

Paul Cropper interview with Cecil Thompson, 1997.

Case 173. 1940s. Petroy Plateau (now in New England National Park) NSW

Albert Mowle told family members that, while he was prospecting for gold, hairy, long-armed gorilla-like creatures entered his campsite to forage for scraps. One, that appeared to be female, kicked a kerosene tin around. Michael Mowle, who passed the story on to journalist Patricia Riggs, said that although his great uncle was fond of telling tall tales, testimony of other old bushmen seemed to corroborate the "gorilla" story.

"Something", he wrote, "terrified the horses belonging to several chaps who were mustering cattle nearby ... their normally docile animals bolted, maddened by fear, and were found ... days later at Georges Creek. One highly experienced bushman said that the only other time he'd seen horses act in this manner was when they had caught the smell of wild animals from a circus."

Macleay Argus, 21 Oct 1976. Credit: Patricia Riggs.

Case 174. 1940 or 1941. Near Eukey, QLD

About six years after Cecil Thompson's yowie encounter (Case 162) his younger sister Leila saw a similar creature:

> In about 1940 or '41, when she was about seven or eight years old … she had a cubby house of sticks and leaves up on the hill on my dad's farm. One Sunday afternoon … she came home white-faced [and] ran and told her mummy, "There's a man up there in my cubby house!" Later she saw a picture of a baboon

in a magazine and said, "That is just like the man in the cubby – sitting down, he was."

The Stanthorpe Border Post, Nov 1, (probably 1992); Paul Cropper interview with Cecil Thompson, 1997.

Case 175. 1946. Near Wilcannia, western NSW

George Nott. (Martin McAdoo).

As soon as George Nott and his family moved into a remote, long-abandoned homestead, strange things began to happen. Huge, five-toed footprints appeared, horses became seriously spooked and then a "bloody big gorilla or something, about six-foot tall, broad, and sort of brownish fur all over him" repeatedly entered the house. On one occasion it grabbed Mrs. Nott by the neck and seemed intent on dragging her outside.

The family retreated to an out-station, but the hairy horror followed them. Mrs. Nott woke one night to find it standing over the bed. After Mr. Nott chased it outside it stamped around in the darkness, "bellerin' like a bull." Later, one of the daughters saw it in broad daylight on the homestead's veranda.

In addition to the frightening "home invasions", the case featured a couple of very odd elements strongly suggestive of poltergeist activity. (See *Australian Poltergeist*, p. 285)

In thc main homcstcad thc Notts heard noises "like a man walkin' about in the ceiling"; objects flew across rooms, and so many pebbles fell on the roof that they "sounded like a heavy shower of rain."

Martin McAdoo, *"If Only I'd Listened To Grandpa"- Recollections Of The Old Days In The Australian Bush*, Lansdowne Press, Sydney, 1980.

For other "home invasions" involving physical violence, see Cases 189 and 190.

Case 176. Late 1940s. Kempsey, NSW

Teenager Kevin Davis got a very close look at a three-foot-eight-inch tall (1.2 metre) ape-man in a gully near the present site of Kempsey airport. Although it was so short, the creature was extremely broad and heavy looking.

As he stood, staring, from a range of only six or eight paces, Kevin noted the creature's flat face, snub nose, and yellow canines. Its eyes, too, were yellow, and its ears were "droopy."

Its body was covered in three-inch (7.5 cm) hair, but that on its head was shorter. It was standing beside a lilly pilly tree and may have been eating its fruit. One huge, clawed hand, twice the size of a man's, was clutching a branch.

Eventually the weird creature broke and ran, leaving several four-toed tracks.

Paul Cropper interview with Kevin Davis, 1995.

This occurred about halfway between Warneton and Belgrave Falls – where a "gorilla" was twice seen in 1871. (Case 224)

Yellow-coloured eyes have been mentioned on a couple of other occasions. Ranger Percy Window, who stood face-to-face with a huge ape-man near Springbrook, Queensland in 1978, said it had "two big, yellow eyes" (*The Yowie*, pp. 2-3) and the Hairy Man that invaded Sir Henry Parkes' Blue Mountains property in 1886 (Case 72) was said to have "eyeballs surrounded with a yellow rim." Another (or the same?) creature, that lurked around Linden many years ago, was known locally as "old yellow eyes." (*The Yowie*, p. 136)

Yowies' ears are almost always said to be small, or concealed by shaggy head hair, so Kevin's mention of "floppy" ears is very unusual. One or two other witnesses, however, have mentioned large, or protruding, ears. (e.g., Case 171)

Case 177. 1949. The Maarlan Scrub, between Ravenshoe and Millaa Millaa, QLD

Vera Hepple told researcher Meryl Tobin that her two brothers came home one evening "sheet white" and claimed they'd seen a huge hairy man. The seven-foot creature had been watching as they played on the edge of Purcell Brook.

A week later, their dairy cows "stood like statues – rigid – watching something moving through the rainforest."

Email from Meryl Tobin to Paul Cropper, Nov 1, 2002.

Case 178. Early 1950s. North Aramara, 35 km west of Maryborough, QLD

Young Michael Meech encountered a strange animal while searching for cows with Errol and Bevan Johnson. Unusually, 40 milkers had "gone bush" on the highest part of the Johnson property. On sighting the boys, the herd stampeded right past, tails held straight out. The boys' dogs, one of which had the reputation of being a fearless fighter, also turned and ran, hackles up and tails between their legs.

> "Then we saw it. I can only describe it as a wooden wine barrel size object covered in jet-black hair with indistinguishable facial features. It was moving ... very quickly towards us with a peculiar upright lumbering gait and we could hear ... guttural grunting sounds. I'm not sure who was first to reach the paddock some two miles below ... the cows or the dogs, but ... I am sure we boys would not have been far behind.

Witness communication with Strange Nation web site; witness interview with Paul Cropper, 2000. Credit: Rebecca Lang.

Michael and his mates had no idea what they'd seen: "No one to my knowledge had encountered or heard much of the yowie at the time." As we have seen, though, "monkeys or gorillas" invaded Mr. Cavanagh's property, just 35 kilometres south-east of Aramara, about 20 years earlier. (Case 156)

Case 179. 1953. South-west of Kempsey, NSW

Neil Bowen wrote that his brother-in-law, Colin Fuller, then 17 years old, and his mate Joe Wright, had watched a hair-covered bipedal animal moving through a cleared area adjoining thick rainforest. This was in the "Molly Milligan" (Marlo Merrikan Creek) area, 25 kilometres south west of Kempsey.

Letter from Neil Bowen to Paul Cropper and Tony Healy, 1995.

Case 180. 1954. Near Batemans Bay, NSW. Aboriginal lore

Percy Mumbulla (Lee Chittick)

> "A doolagarl is a gorilla-like man. He has long spindly legs. He has a big chest, long arms. His forehead goes back from his eyebrows. His head goes into his shoulders, no neck. They live now on Cockwhy and Pollwombra Mountains.

Percy Mumbulla, quoted by Roland Robinson in "Three Aboriginal Tales", *The Bulletin*, 13 Oct 1954.

Case 181. 1956. Young, NSW

In his memoirs, written in 2007, retired broadcaster Frank Avis vividly recalled a strange encounter he'd experienced as an 18-year-old on a property just outside Young:

> "We were ... rabbit shooting ... [on] the side of a hill ... cleared, but up above remained heavy forest ... All of a sudden there was this tremendous noise, like a train coming from behind us.
>
> Seconds later this huge red kangaroo came crashing past, within a metre or two, scaring the life out of the three of us. We went berserk, firing round after round ... as it disappeared down the hill ... in '55-'56 what did we know? We were just galahs like all the other youths of the time.
>
> Anyway, as the 'roo headed away, for some reason I looked up the hill and there, standing at the edge of the bush, was this creature looking straight at me, eating leaves from a tree, completely unconcerned. My mates were still firing wildly and yelling but this creature wasn't even remotely spooked. It looked almost human, with two arms and legs, but was covered in dirty red-orange hair.
>
> We were about 25 metres away; I'd put its height at around 5 feet .... One thing I remember, and will carry with me, is that it looked straight into my eyes. There was this mesmerising contact. The animal/creature had big, brown, sorrowful eyes. I will never forget them.
>
> I must have said something pretty dramatic and with a certain amount of vigour, because my two mates stopped shooting and looked at me in amazement. I remember explaining how I'd "seen something" turning back to [look] up the hill. "Please God", I begged, and remember I'm an atheist, "Please God, let it still

be there." IT WASN'T ... all that could be seen was the rustle of the leaves.

I then faced a momentous decision. Did I [risk] being labelled an idiot, or did I lie and try to change the subject ... I chose the latter. An 18-year-old in '56 was about as sophisticated as a 12-year-old in 2007. I had absolutely no benchmark in making my decision. There were no terms of reference. ... So I blurted out something stupid and we continued on ... with both mates looking at me from time to time, wondering what the hell I was on about.

Frank Avis

[Later, when the landowner asked about their excursion, one of the boys said], "Oh, and Frank saw something up there in the bush" and burst out laughing. Farmer Jim looked across and asked what it was ... I remember making some stupid excuse, like it might have been a cow or something. I looked up and Jim's eyes met mine. My jaw dropped. The subject was never mentioned again, even though Jim and I met several times, but I knew the minute he looked at me that afternoon. The thing I'd seen, he'd seen it too. I confess that made things a lot better.

Frank Avis, *Memoirs of 42 Years in Radio* (2007). http://www.frankavis.com/blog/default.asp?id=268.
Credit: Jon Downes and Rebecca Lang.

Strange daze

It's interesting to note that the oddest detail in Frank's story – the "mesmerising" effect of the creature's "big, brown, sorrowful eyes" – has featured in quite a few other reports. Readers may recall that Mamie Mason (Case 148) said that the Burnt Bridge yowie "sort of had me hypnotised." Several modern era witnesses have suffered the same mind-numbing effect. (e.g., Cases 183 and 214)

Case 182. October 1957. 12 km south of Gympie, QLD

Researcher Steve Rushton interviewed J.D. McDonald, of Cooroy, on 17 Feb 1994.

Mr. McDonald, his wife and his father-in-law encountered a huge man-like creature, at night, on the old highway. More than eight feet tall, it was covered in whitish hair "like a polar bear," and seemed to have large, flat feet. It emerged from scrubland and ran with astounding speed across the road in front of their car.

Case 183. c. 1958-1962. Near Batemans Bay, NSW

Sawmill owner Laurie Allard is a lifelong resident and respected citizen of Batemans Bay. When they were in their early twenties, he and a mate, Bill Taylor, often went wallaby shooting. One day, not far from Cullendulla, they encountered a yowie.

To avoid accidents, they would often stand on tree stumps about 200 yards apart, sending the dogs ahead to flush out the game.

> "There was a gully about 150 yards in front of us; the ground was pretty clear, I looked down and there was this big grey thing running up the gully, and I thought it was my mate. I knew that didn't add up, but we were the only people in the area, so [I took] it for granted that it must be him, running to a new position. It took really large strides and I thought, "Boy, he's really running today!" Later I realised it was running *too* fast; a man could not run up that gully at that speed.
>
> Anyway, it ran around the head of the gully, took a turn and came back to where I was standing on my stump [and] stopped about 60 or 70 feet away ... And it had a wallaby over its shoulder [holding the black tail] in front of its chest with both hands. It was no taller than about five foot six [168 cm] and grey all over, from head to foot. I couldn't say if it was long hair or short hair, but it was grey, about the same colour as the clothes Bill used to wear ... It stood looking straight at me.

Laurie acknowledges that the shock of meeting such an odd creature face-to-face sent him into a state of confusion. Although it obviously wasn't another human or a kangaroo, his mind simply would not accept what his eyes were seeing.

> "Funny thing: it gave me a really hairy, scary, feeling. It was stockier than a man and shorter in the neck, but I didn't really absorb it. And I thought it was my mate – I still had it half in mind it was him – and I spoke to it: "What have you got – have you got a wallaby, Bill?"
>
> After watching me for maybe a minute he turned around and ran away into the scrub. Then Bill came over to me and said, "What were you doing running around in the bush?" and – I was still confused – I said, "What were *you* doing

running around?" He hadn't seen this thing but he'd heard it charging around – and he had the same scared feeling. All the dogs came in, and they were that terrified that they were in amongst our legs – they *would not* leave us. We called it a day and left.

Although he still works in many remote corners of the south coast hinterland, Laurie has never so much as glimpsed another yowie. "I often think of them, though," he says, "and I feel really privileged to be one of the few people to have seen one up close."

Witness interviews with Tony Healy, 1998 and 2003.

Case 184. September 1959. Byng, 16 km south-east of Orange, NSW

Here's one for the sceptics – a well-organised hoax that apparently fooled some people for quite a long time.

Throughout the entire episode, however, the terms "yowie" or "Hairy Man" were never used. The "creature" in question was always known as the "Byng Bunyip."

Freddy Nunn meets the 'Bunyip' face-to-face

It began on the night of September 3, 1959, when Max Spicer of Byng saw something standing in the middle of the road near the cemetery. Stopping his car, he saw a creature that seemed to be about three and a half feet tall. It was covered with long, black hair, had glaring red eyes and arms that reached to the ground. It scampered into the graveyard.

After other locals and three Sydney men encountered the same thing on the same road, armed parties began scouring the area and the story received front-page coverage in the national press. One night a truckload of hunters spotted the "creature" but when one of them accidentally doused the spotlight it vanished into the shadows. It was supposedly encountered several more times but always escaped. After a while the sightings became fewer, and the "Bunyip" seemed to just fade away.

In 2002, however, our colleague Roger Frankenberg uncovered the awful truth. The "Bunyip" had been a pint-sized teenage panel-beater, Graham "Darby" Offen. The hoax had been dreamed up by some of his mates, and, as Darby said later, "they only picked me to be the bunyip because I was a little bugger."

One of the merry pranksters, Denis Gregory, told Roger that they'd dressed Darby in a hairy hessian suit. His "big red eyes" were reflectors, as used on bicycles. He would wait in the middle of the road and, when a vehicle approached, would jump around a little before scooting away to where his mates had a car hidden with its motor running.

One night the prank almost got Darby killed – "the small hole I looked through in the suit got stuck on my nose and I couldn't see a thing. I could feel the lights burning into me and I didn't know which way to run … my mates were yelling to me to run, but I didn't know where to and I thought, 'I'm a bloody goner this time.' But I got off … fell down a culvert and me mates grabbed me. That was too close for comfort, so I decided to retire … there was no way I was going to get my head blown off."

Although he'd taken part in some of the "Bunyip" hunts and actually sighted the hessian-clad horror, one local man, Freddy Nunn, was supposedly not let in on the secret until about 1990. If that detail is true, then the "Byng Bunyip" is one yowie-type hoax that fooled at least one man for almost 40 years.

Unreferenced Sydney tabloids, 29 and 30 Dec 1959; *The Picture*, 25 Jul 1989; Denis Gregory and Alf Manciagli, *There's Some Bloody Funny People on the Road to Broken Hill*, 1993. Credit: Roger Frankenberg.

Case 185. Summer 1962. Near Mount Buller, VIC

A Knoxfield (Melbourne) man wrote to yowie witness Maria Speer (*The Yowie*, pp. 270-71) to say he'd glimpsed a large, hair-covered creature in Howqua Valley, near Mount Bulla in 1962. It had been moving away through dense bush. Although he didn't see its legs, he was sure it was "too thick above shoulders (sic) to be a kangaroo."

Letter from witness to Maria Speer, early 2000.

Case 186. 1963. 14 km south of Tenterfield, NSW

Elsie Mitchell, whose husband Bob saw two yowies in the QLD/NSW Border Ranges in 1928 (Case 146) saw three or four strange creatures while driving along an old bullock track to the south of Tenterfield:

> "It was dusk and there were lots of trees along the side of the track. At one point I noticed what I thought were three or four trees cut off about six feet [183 cm] from the ground. I remember thinking it was a funny way to lop trees – then they moved.

There was also a strong smell like rotting flesh and a low rumbling noise. I was a bit scared because I didn't know what it was. Some years later I read a report about a yowie sighting and it mentioned that distinctive smell. That convinced me I had seen yowies.

Sunday Mail (Brisbane), 9 Nov 1980.

Case 187. 1964 or '65. Wyandra, 95 km south of Charleville, QLD

Soon after seeing a strange, hairy, man-like creature crossing a road near Wyandra, a local postman supposedly quit his job and left the area. Other locals, including Frank Colgin and Mrs. Summerfield, reported finding several sets of very weird-looking footprints.

According to the *Sydney Sun* the tracks were "almost eight inches by six inches [20 X 15cm] … they had three front pads – two with three toes and one with two toes. There was also a rear pad, or it might have been a heel …"

Despite the supposedly multi-toed footprints, it was theorised that a large ape-like creature was on the loose.

The Sun (Sydney), 1 Aug 1965.

Case 188. Between 1965 and 1968. Drummer Mountain, VIC

The incident occurred as an Aboriginal family was driving home to Bermagui from Victoria on the Pacific Highway. The youngest child, Colin Andy, was then about 10 or 12 years old.

Late at night they reached the top of Drummer Mountain, about 20 kilometres east of Cann River, and stopped for a toilet break. As the adults walked into the bush, Col strolled down the road behind the car. There wasn't a house light or campfire to be seen, but the road was lit up brilliantly by a full moon.

Presently, a strange creature walked out of the scrub 80 metres away and crossed the road. Though bipedal, it was not human; its arms were so long that its hands almost touched the ground, and it "walked hunched over – like an ape."

Because he was born into the Yuin tribe, Col was quite familiar with the *dulagar* legend, but for a reason he is still at a loss to explain, he didn't feel like telling anyone what he had seen. When he finally confided in his mother in the mid-1980s, she agreed he'd almost certainly seen a *dulagar*.

Witness interview with Tony Healy, 27 March 1998.

Case 189. 1965 or '66. Mt. King Billy, near Mt. Buller, VIC

When cattleman John Lovick was about 19 or 20, he accompanied his father Jack and four or five other people on a fishing expedition into the high country. They bivouacked at King Billy Hut; a one-room shack in remote, trackless country. As there were only two double bunks, most of the party had to sleep on the floor. After "lights out" the hut was in darkness apart from moonlight shining through the door, which was wide open. Although John had scored one of the top bunks, he didn't have a particularly restful night.

At around midnight he was woken by "this bloody heavy thing pressing on my chest. It had its hands around my throat, trying to strangle me. It had a strong animal smell like you'd run across in a zoo. Its hair felt coarse and I could feel its hot breath in my face. I yelled and screamed and tried to throw it off, and we fell onto the floor."

The creature broke loose and although John didn't witness it himself, "a couple of blokes saw its silhouette as it walked [upright] out the door. It was so broad it filled the doorway, but they said it wasn't very tall, five foot at most.

The dogs were making a bloody ruckus and the old man went out and tried to sool them onto [the creature] but they wouldn't go out more than five or 10 metres: they kept coming back with their bristles up."

In the morning they could find no tracks on the hard, dry ground, but there were bruises on John's neck – such as might have been left by strong fingers. He speculated that the creature had been used to sleeping in the hut, which was unattended for months on end, and became angry when it found the place full of snoring interlopers. But he still wonders why it should have stepped around the men on the floor and attacked him in the bunk farthest from the door.

Witness interviews with Bernie Mace and Pam Bryant, c. 1990, and with Tony Healy 23 May 2005.

It's interesting to compare John's ordeal with that endured a couple of years later, and hundreds of kilometres to the north, by sawmiller George Gray. (Case 190, below)

Yowies have been reported near Mt. King Billy on other occasions. Howqua Valley, where one was seen in 1962 (Case 185) is only a few kilometres to the west, and cryptozoologist Bernie Mace watched through binoculars as an eight to ten-foot-tall specimen walked down a slope near the mountain in April 1988. (*The Yowie*, p. 263)

Case 190. September 1968. Kookaburra, 50 km west of Kempsey, NSW

In 1976, retired timber worker George Gray told journalist Patricia Riggs about a terrifying encounter he'd had with an apparent yowie eight years earlier.

At the time, Mr. Gray, then 55 years old, was working at Kookaburra, an isolated saw-milling settlement on the Carrai Plateau, where he camped in a hut surrounded by dense scrub.

On the night in question, he woke to find he was being choked and shaken by a small but powerful creature, covered with grey, bristly hair. Although small, it was very broad, "like a well-built little man" with a short, bullish neck and large legs. Its arms were short, yet strong enough to throw him around with ease.

With its hands around his throat, it seemed intent on dragging him out of the house. Nightmarishly, as he fought for his life, the terrified timber worker was unable to get a good hold on his assailant's skin, because it felt loose and slippery.

Mr. Gray defended himself for ten desperate minutes until his assailant abruptly released him and ran out the door.

His two young sons, who happened to be visiting, had been awaked by the ruckus, but were afraid to intervene.

Macleay Argus, 4 and 18 Sept 1976; *Sydney Sun Herald*, 12 Sept 1976. Credit: Patricia Riggs.

Mr. Gray's tussle with the dastardly dwarf was certainly nightmarish, but before sceptics dismiss it as just that – a very bad dream – we suggest they look again at Cases 175 and 189, which detail very similar nocturnal attacks.

Case 191. 7 August 1970. Near Katoomba, Blue Mountains, NSW

This item requires quite a bit of commentary because it features the testimony of a yowie researcher. Not just any old yowie researcher, in fact, but Rex Gilroy – the man who advanced the public's awareness of the phenomenon to such an extent that we now think of 1975 – the year he began appealing for eyewitness reports in the press – as the beginning of the modern era of yowie research.

Rex often claims to be "the father of Australian yowie research", and no fair-minded person would deny him that. During the mid-1970s he almost single-handedly dragged the hairy giants out of obscurity and into the consciousness of the (admittedly still sceptical) Australian public.

By pointing out the widespread Aboriginal Hairy Man traditions and the numerous sightings by British colonists, he did a great deal to show that the yowie phenomenon is more than just a myth. If only he'd left it at that …

Unfortunately, however, he went on to make a long series of increasingly outrageous claims, causing many of his fellow researchers, not to mention journalists, scientists, and casual observers, to remain cautiously sceptical of anything he said.

There was the matter of the "fossilised *Gigantopithecus* footprint", a cast of which he displayed for several years. He claimed to have found the "footprint" near Kempsey, NSW.

Remains of *Gigantopithecus*, a huge hominid that has been extinct for about 500,000 years, have been found only in southern China, Vietnam, and northern India. As those remains consist only of jawbones and teeth, it is very difficult, for even the most credulous, to accept his assertion that what he displayed was a footprint of one of the creatures.

Seemingly unwilling to accept the notion that extraordinary claims require extraordinary evidence, he also wrote about a UFO base nestling under the Blue Mountains, "Egyptian pyramids" he examined in Queensland, "Phoenician inscriptions" he deciphered in NSW and much, much more. If that wasn't enough to set the alarm bells ringing, he also modestly admitted to being to be one of a select band of "extraordinary geniuses" chosen and guided by space aliens.

Despite all that, Rex is basically right – about the yowie, at least – and the sceptics are wrong. The yowie mystery *is*, as he always maintained, a genuine phenomenon worthy of serious attention.

But as for his own "sighting" …

In 1975, when he began appealing for information about yowies, he announced that he'd seen one himself, five years earlier, below the Ruined Castle rock formation, four kilometres south of Katoomba. Since then, several differing versions of the story, most of them containing statements attributed to him, have appeared in magazines and newspapers. He accounts for the contradictions by saying he has been extensively misquoted. Unfortunately, however, that isn't entirely true – several of the contradictions appear in his own books, articles, and letters.

In June 1977, he wrote that the hair-covered, "ape-like" creature appeared to be four to five feet tall, moved swiftly and disappeared "within a few seconds." In 1980 he wrote that it stood five to six feet tall. Again, it "disappeared within seconds." He added that it was 40 yards away, and that, as he only saw it from behind, he "never caught sight of the creature's face." In 2001, he remembered it being 15 metres away, and added that it was male, that it had "big eyebrows" and long, trailing head hair. He watched for "four to five minutes" as, oblivious to his presence, it moved slowly across a slope, carrying what looked like a digging stick, apparently fossicking for roots.

Despite having previously written that he never saw the creature's face, Rex presented, in his book *Giants From The Dreamtime*, a very detailed close-up sketch of it, complete with nostrils, teeth, ears, and eyes. The sketch bears a remarkable resemblance to an artist's impression of a *Paranthropus* face that appeared in a 1966 Time-Life book on the prehistory of mankind.

Rex Gilroy, "Gorilla Giants at Katoomba", *Psychic Australian*, June 1977; Rex Gilroy, "Why Yowies are Fair Dinkum", *Australasian Post*, August 7, 1980; Rex Gilroy, *Giants From The Dreamtime*, pp. 173 and 189; Rex Gilroy, *Australian UFOs Through The Window Of Time*, URU Publishing, Katoomba, NSW, 2004, pp. 257-58; *Early Man*, Time-Life Books, 1966, p. 57.

Case 192. 1971. The Jingeras, south of Captains Flat, NSW

In the winter of 1971, as 22-year-old Jim Banks and his mate Stan Hunt were spotlighting on Wild Cattle Flat Road, they noticed an unusual pair of eyes. Reflecting red, and seemingly very high off the ground, they were between the road and the boundary fence, some distance ahead. Something, it seemed, was peering around a tree at them. "We said, 'What's that?' It wasn't a kangaroo."

Driving forward, they saw a large creature about 50 metres away on the other side of the fence, running across a cleared paddock that rose sharply towards the edge of the bush. It seemed to be over seven feet [2.13 m] tall and was covered with "dark brown to grey hair … maybe four to six inches [10-15 cm] long". It was tail-less and its neck was either very short or non-existent. Its head was down and its shoulders were hunched as it "plodded away, like a big heavy front row … the gait didn't correspond with anything I'd ever seen before … it was clumping, like it was very heavy."

This might upset some readers: Because he was sure the creature wasn't human, Jim decided to fire "to see how it reacted." It was an easy target: "I couldn't miss." Raising his .22 magnum semi-automatic, he pumped two bullets into its broad back. Although he had a ten-shot magazine, he immediately stopped firing because of the creature's startling reaction. Throwing up its arms, it let out a scream like nothing the men had ever heard – wild and totally unnerving – "an unearthly sort of squeal – real high-pitched." The creature kept running. The men called it a night, went home, and never hunted that stretch of road again.

At the time, neither Jim nor Stan had heard the slightest thing about the yowie legend. They had no idea that the site of their encounter was in the heart of an area – The Jingeras – that was, in colonial days, a notorious haunt of the Hairy Man. (e.g., Cases 35 and 75)

Jim Banks interview with Tony Healy, 1993.

Case 193. 1972. Cullendulla, NSW

Early one morning, George Birch of Bowral was driving from Nowra to Bega on the Princes Highway. Just on daylight, he stopped his truck near Cullendulla, six kilometres north of Batemans Bay, to check the load. As he was doing so, a strange "screeching" noise came from the forest on the western side.

Walking to the embankment and looking down into a gully, he was astonished to find two strange animals staring back at him from a small clearing about 20 metres away. Although they were covered in brown-to-black hair that was rather short, "not wild-looking or untidy," his impression was of two hairy people rather than apes. But they had flat noses and arms that were longer than those of humans. Their posture was rather stooped and one appeared to have breasts.

Although they were shorter than George (who is six feet three inches tall), he cheerfully confesses they gave him a terrible fright – he is certain his hair actually stood on end. Quickly retreating and glancing back, he saw the animals walking away through the bush. In his haste he almost ran into the front of the truck and, once inside, sat for ten minutes before regaining his composure.

He had no way of knowing it, but his encounter occurred within two kilometres of where Laurie Allard came face-to-face with a five-foot tall ape-man about ten years earlier. (Case 183)

Witness interview with Paul Cropper, 29 Jun 1993.

Case 194. 1973. Middle Brother Mountain, near Taree, NSW

Although the following occurred when Eric Bond, now middle-aged, was only eight years old, the details remain etched in his mind.

He contacted AYR in March 2017 and was interviewed by Paul Cropper on 6 Apr 2017.

> "I was up at Middle Brother Mountain on a school excursion … about 25 of us – my mum, the nuns, and a bunch of other people.
>
> At about 11 am, me, Len [his best mate], the two Latham brothers and Jeremy White – wandered off, down towards the creek. After a while there was a movement in the bush – following us, moving parallel. We thought one of the older kids was trying to spook us, so we started throwing rocks at it and laughing – and it broke cover about 65 metres away – running away. It was big and hairy and scary! [Laughs]
>
> We saw it for maybe two or three minutes, because he had to go up the gully ahead, along the side of the ravine, over the top and into the distance. He was

pretty solid – a big boy – probably 160, 170 kilos … 7½ to 8 feet … stooped over a little bit – its shoulders were rolled forward. It looked well-muscled. We could see arms and legs. We saw it side-on; never saw the front. It looked humanoid – on two legs the whole time, and had a really big stride – I reckon about a metre and a half. It didn't walk like we walk – it sort-of loped, sprang – big steps. It moved pretty quick.

There was fur all over its body … a bit matted, mostly dark brown with lighter patches – mottled … reasonably long and stringy – longer than, say, 3 to 4 inches.

I was rooted to the spot … couldn't move, absolutely terrified. We just stood and watched it – didn't say a word. I wouldn't go out after dark for about three years. It scared the hell out of me!

We told our parents and friends. The rangers laughed and said we were imagining things, and our parents told us we shouldn't be making things up. The kids at school gave us a hard time for about two years!

The Latham boys now deny seeing anything, but me and Len stand by what we saw. He's now in Tasmania. I saw him recently and we talked about it.

It's interesting that while struggling to describe the seemingly half-man/half-ape creature, Eric, like many of our other eyewitnesses, referred to the yowie as both "it" and "him."

Worth noting, also, is that 60 years before his encounter, another yowie was seen at Kendall, just 10 kilometres north of Middle Brother Mountain (Case 133) and that the Taree-Wingham area, just to the south, has produced more than a dozen reports since the early 1970s. (*Out of the Shadows*, pp. 126-29 and *The Yowie*, pp. 237-39)

Case 195. c. 1973. Near Bexhill, 6 km north of Lismore, NSW

After contacting AYR, Mark Pope was interviewed by Paul Cropper on 30 Jan 2000.

While camping beside a creek, Mark and two other teenagers were visited by – and shot at – a huge, hairy, man-like creature.

"The tent was open at the front, so we had a clear view out. That's when we saw it … just standing there … 20, 30 metres away, in a clearing. There was light behind it … we could see him silhouetted. It was facing us, so you couldn't see any facial features. The face was dark … matted hair … it was probably close to eight feet high. We could see its hands quite clearly. [Its arms] were almost to its knees. The head seemed to be just sitting on the shoulders. It seemed to be a bit pointed; it wasn't rounded.

We were just dumbfounded; silent and frozen. It was probably less than a minute, then it just slowly moved off ... but at a tangent, so, if anything, it got a bit closer. As it went up the side of the tent it was out of view. It never did anything to threaten us, but we were kids and it was pretty big. Well, one of the guys picked up a shotgun and fired it through the side of the tent, which was more or less where it would have been … but we never heard noise out of it, but you could hear it moving off quicker than it had been. So I don't think it got hit but, certainly, I think it was startled.

Twenty-four years later Mark encountered a one-metre-tall, ape-like creature – possibly a junjudee – near Woodenbong NSW. (*The Yowie*, p. 125)

Case 196. August 1973. Burleigh Heads, QLD

As teenagers Alan Livingstone, Brian Morey, Ken Kilner and Gary Hoffchild were driving along Tallebudgera Creek Road, at about 11o'clock one night, they saw what looked like a pair of glowing eyes in the scrub and stopped to investigate:

> "As soon as we got near the thing, it came charging out towards us. It was about five foot eight inches tall, had two arms, walked on two legs like a man, and was covered in reddish fur. We didn't stay around too long to find out exactly what it was … we took off like rabbits. As we ran back to the car, we could hear its padded feet thumping the ground behind us.

In the early 1970s the Gold Coast stretched more than 40 kilometres north to south, but the greater part of it extended inland just a few hundred metres. So the spot where the boys were monstered by the ape-man was beyond the urban strip, on the edge of some pretty rough scrub.

After the boys, who'd never heard of yowies, approached local zoos to ask if a bear had escaped, the *Gold Coast Bulletin* got wind of the story.

Probably because it occurred at the tail end of the early modern era, when awareness of the yowie phenomenon was at an all-time low, the journalists, like the baffled teenagers, didn't know what to make of the incident. As a result, they labelled the creature the "Burleigh Bunyip."

Gold Coast Bulletin, undated clipping; Brisbane *Sunday Mail*, 5 Aug 1973;
Sydney *Sunday Telegraph*, 5 Aug 1973.

Case 197. Winter 1974 or '75. Killawarra, NSW

One sunny afternoon Alwyn Richards and his sister rode their horses up a hillside near Alwyn's property. Ahead was a narrow strip of forest, then a cleared firebreak, and beyond that a large area of rough scrub.

The horses normally loved to charge up the slope, but on this occasion they shied violently away from the wooded area. Alwyn eventually dismounted and led his reluctant steed through the trees with his sister following. On the other side they were amazed to find a huge, shaggy creature standing in the firebreak.

It was nine to ten feet [about 3 metres] tall, broad chested, very muscular, well-proportioned and covered in long, untidy looking hair. Although its arms were longer than those of a man, it appeared more human than ape-like. Parts of its face not obscured by hair appeared to be black.

It stood staring at them for what seemed like several minutes as Alwyn approached to within about 30 metres. A "terrible burning smell" was evident. The huge creature finally turned and walked away, moving surprisingly quickly, and stepped right over a four-foot [1.2 metre] fence without breaking stride.

Alwyn Richards. The yowie stepped over this fence without breaking stride.

Alwyn walked straight over to where it had been standing. No clear tracks were evident, but plants were squashed, and the ground was distinctly warm, "like where a cow has been lying down".

Alwyn Richards, interviews with Geoff Nelson, Paul Cropper and Tony Healy, June 1993.

Case 198. c. 1974. Near Montville, QLD

The primary witness, Mark McDonald, of Cooroy, was interviewed by our friend Steve Rushton in 1994.

As Mark and a mate drove slowly along the foggy Montville to Maleny road, a huge hair-covered ape-man jumped from the embankment and ran across in front of them. The incident disturbed Mark so much that, even 20 years later, he could hardly bear to talk about it.

In view of the several other instances where yowie sightings appear to "run in the family," we'd be remiss in not mentioning that Mark's parents and grandfather had a similar experience near Gympie in 1957. (Case 182)

Case 199. c. 1974. Tasmania

These stories came from a resident of the village of Cooee, Tasmania, who, after hearing of Maria Speer's yowie sighting near Eden NSW (*The Yowie*, pp. 270-71), wrote to her in 2000 to relate a couple of similar events from the island state:

> "My daughter, who was about 5 yrs old, saw in dense scrub, something she described as a big ugly hairy man, and was very frightened … We never saw it ourselves but after she calmed down she said it had been standing on a log watching us, but by the time she drew our attention to it, all we saw was the thick bushes moving and the sound of something large & heavy going through the scrub.
>
> "Nothing was said about this to anyone … But about 2 mths later two men who were logging in that same area, told us they saw, well, to use their very words, … the weirdest thing they had ever saw in the bush in their years of logging. They were loading logs, it was raining quite heavily and windy, they almost had the truck loaded when the other chap said Look! What the hell is that, and on the muddy road not far from them stood this ape like man…" (Second page of letter missing).

In 2006 we documented, as best we could, two other yowie reports from Tasmania – the only ones from the island state then in our files. (*The Yowie*, pp. 233 and 261)

Maybe our research has been inadequate, but since then we have gathered only two more – the interesting, though incomplete, story above, and a much more detailed, first-hand report shared with us by Dean Harris, of Smithton. (Case 217)

The extreme scarcity of Tasmanian reports presents a problem not only for the "flesh-and-blood" school of yowie enthusiasts, but also for sceptics.

If the creatures are flesh and blood, their shaggy ancestors could easily have walked to Tasmania before it was cut off from the mainland 12,000 years ago – so we could reasonably expect to hear of many other sightings there. Similarly, even if the entire yowie phenomenon is just an amalgam of hoaxes, hysteria and hallucination, as sceptics believe, there is no particular reason why they shouldn't be reported as often in Tasmania as in every mainland state.

The situation isn't helped by the extreme rarity (as far as we know, at least) of references in Tasmanian Aboriginal lore to beings that may, or may not, be yowies.

During the terrible struggle known as The Black War of 1824-31, native Tasmanians suffered devastating casualties. In the following decades, disease and ill-treatment reduced their numbers even further, so that much of their culture was lost to history.

Throughout those dark days, however, a handful of British colonists took some interest in their captives' traditional beliefs, and a few tantalising fragments of lore that *might* refer to yowie-like creatures were documented.

During the Black War, the Tasmanians put up a very good fight, but had one fatal tactical weakness – strangely enough, they would not attack at night.

According to George Augustus Robinson, Jorgen Jorgenson and others deeply involved in the war, the Aborigines were afraid to leave their fires because of "an evil spirit they believed lurked in the dark" – they "were always fearful of travelling by night, [and] never did so, until they were pressed hard by the parties in pursuit of them."

One man, Woorrady, told Robinson that the fearsome entity was, "like a black man only very big and ugly, and that he travels like the wind, that he comes and watches the natives all night and before daylight he goes away."

Our sources are sketchy, but it seems the midnight rambler was known as *Rangeowrapper*

Apart from it being "very big and ugly," man-like and remarkably quick on its feet, there is nothing to indicate that it was otherwise yowie-like – there is no mention, for instance, of the demon being hair-covered or given to frightening vocalisations.

There is, however, one other fragment of Aboriginal lore that might suggest the bogeyman was indeed a cousin of the mainland yowie.

After writing, sceptically, about mainland *yahoo* (yowie) beliefs, Louisa Anne Meredith (Case 12) mentioned, in passing, that the term yahoo, "... being used to express a bad spirit, or 'Bugaboo', was common also with the aborigines of Van Diemen's Land."

Case 200. November 1974. Blackheath, NSW

While cycling near Blackheath aerodrome, a 12-year-old boy saw a large creature running through scrub about 50 metres away. More like a gorilla than anything else, and extremely muscular, it was covered in short brown hair with some patches of skin visible. Its fingers were long and held in a hooked position. Its large head was more densely hair-covered than the body. No ears were visible.

Witness interviews with Paul Cropper and Tony Healy, 1981 and 1982.

Case 201. Summer 1974. Between Casino and Whiporie, NSW

Having attended the Lismore Speedway, Michael Allison and his two brothers began driving home to Grafton. They took Summerland Way, a road they knew well. The night was perfectly clear.

Around 1am, as their vehicle negotiated a sweeping left-hand bend, its powerful driving lights caught a huge ape-like animal about 200 feet ahead, almost directly on the centre line of the road. It turned its head to face them and, apparently dazzled, stood stock-still, which was just as well, because the speedway fans were travelling at around 100 miles per hour. They passed, Michael estimates, within one foot (30 cm) of the gob-smacked gorilla. Seconds later, several hundred metres down the road, the brothers turned to each other and said as one, "Did you see *that*?"

The animal was "enormous" – about three feet wide at the shoulders – and although "hunched over" was much higher than the roof of the car – "at least eight feet." It was covered in long, dark brown hair. The shoulders and face reminded Michael of an ape or gorilla, but he also used the term "Neanderthal" several times.

Grafton Examiner, undated article, Apr 1990.
Witness interview with Paul Cropper, 22 Dec, 1996.

A close shave: the Allison brothers' yowie

CHAPTER THREE
THE MODERN ERA, 1975 – PRESENT

It would take a book half as big as the Sydney telephone directory to document all of the modern-era reports now in our files, so, as mentioned in the introduction, we will present only a "judicious selection" of post-1974 cases.

All the cases we've selected came to our attention after our previous book, *The Yowie*, went to press in 2006, so they will be new to most readers.

To provide a chronological spread of modern cases, we've included at least two or three reports per decade. We've also aimed for a geographical spread, to illustrate the continent-wide nature of the phenomenon.

Priority has also been given to cases that align with Aboriginal yowie lore in some way, to those that spring from areas noted for Hairy Man activity in the colonial or early modern eras, and to those that contain odd, but possibly significant, details.

Some stories were included because the witnesses provided unusually vivid word-pictures of their experience, others because they were shared by people whose honesty and good character seemed beyond doubt, and others simply because the witness had a particularly amusing turn of phrase.

Case 202. 1975. Blue Mountains NSW

Mark Franklin reported this incident to AYR on 27 March 2015 and was interviewed by Paul Cropper three days later.

Grose Valley, Blue Mountains

Mark and his mates grew up riding motorcycles through the mountains, often along dirt tracks to an area called The Quarries.

> **Mark:** "This happened when we were around 14 years old. One mate had a brand-new motorcycle. We'd just started down the Catchment Road and we saw a cow that had been just torn apart – mutilated very, very badly – and we thought, 'Wow – this is strange!'
>
> "Then, about a kilometre before The Quarries – in the middle of nowhere – we came across *another* cow, and this one was also badly torn up –pretty horrific.
>
> "We had our bikes packed with gear … It was one of the first times our parents had let us go camping by ourselves. We picked a spot between both of the quarries and early in the evening we started hearing this noise – like a deep, growling *howl* – something we'd never heard before … it sounded like it was way over – where we saw the second cow.
>
> "We'd set up camp, had a big fire, and the bikes were nice and close. And we heard the roar again, a bit closer, and we were trying to come to terms with what this thing could be … [It was] a deep, throaty roar – a husky roar.
>
> "The sound kept getting closer and closer; then we got this real strong odour, and it was *foul* – it was like, *mouldy*, *wet* – like hair, terrible … really, really nasty – it was on the wrong side of wrong. And we're thinking 'Okay …', getting a bit edgy.
>
> "We threw more wood on the fire, it was … raging, and lit up the area quite well. The night was just incredibly still. It was a dark night – stars but no moon.
>
> "The next time we heard this roar it seemed about 100 metres behind the tent, in some very, very thick scrub that you'd have a lot of problems trying to walk through. The smell was still strong, and now we could hear twigs and branches rattling and snapping – something coming through.
>
> "When it was within about, I reckon, 10 metres from the edge of the scrub behind the tent, one of my mates started running [leaving his brand-new motorcycle behind]. So my other mate and I ran to our bikes. As we were starting off, we stopped and looked back into the firelight and saw this big, dark figure that had pushed the branches over.
>
> "It wasn't shaped like a person – just a big, *big*, human-like shape, I suppose – because it had a head. Long arms … it looked really hairy, and it was huge – it must have been bigger than, gee, seven to eight feet tall. *Huge* – illuminated by the fire. It was big and solid … its legs were big and *thick*. And it was just standing there.
>
> "We thought, "We're not hanging around to get up close and personal with

it." So we took off, and my mate stopped and picked up the other kid and we drove out to Katoomba – eight kilometres on a dirt road in the middle of the night. We'd left everything [including the new motorcycle] there.

The whole episode from beginning to end lasted around 45 minutes. This is giving me goose bumps again – I can't believe it!

"When we got to my mate's place his big brother and some of his mates wanted to drive out and have a look – but we weren't particularly keen. We reluctantly got into their car and went back out, and when we got there, there was no smell [and no creature, but] the tent was trashed: just torn to pieces. We then drove around the top of the quarries but didn't see anything.

"Next day, we went out and picked up the motorcycle and walked through one of the quarries and we found these huge footprints in the wet clay, about 50 metres or so from the campsite.

"Around two or three days later, Rex Gilroy got the story off one of my mates, knocked on my parent's door and had a chat with me. He later rang [and said] he'd gone out there and taken plaster casts of the footprints.

"[There were] about three tracks, two were very distinct – and they were *big*. They didn't even look like a human footprint … a round heel at the back. Big and flat and very broad, from memory, with a lot of definition with the toes. Five toes, from memory, like a person's, but they seemed small and fat.

"The bush that it came through – you just *can't* walk through there – it's so dense and thick – it would really scratch you. No one would go to that trouble [to perpetrate a hoax].

"About five years later, we started to drive out there again. But as we drove, and talked about what had happened, we [got scared] and turned around – we didn't make it! Just talking about it now, I've got goose bumps – it still puts chills up my spine – and I'm now 53! It's something I'll never forget: we had an encounter with the Australian bigfoot – a yowie.

"Years later, I tried to take the wife and daughters out there but couldn't do it – I can't even drive out there in the middle of the day!"

The site of Mark's encounter is at the very end of North Road, about 10 to 15 km west, as the crow flies, of Minnehaha Falls. The area is no longer as wild as it once was – in recent years one quarry was converted into a motocross track.

Case 203. May/June 1978. Llangothlin, 10 km north of Guyra, NSW

Neil Watterson reported this remarkable incident to AYR on 3 Oct 2017 and was interviewed by Tony Healy on 2 Dec 2017.

Neil: "Back in 1978 [while heading for Queensland, on holiday from his NSW Railways job] with a friend of mine, Michael, I drove up through Tamworth, Armidale, and Guyra, up to a little town called Llangothlin.

"I was driving a 1976 VC Valiant. It had a truck engine in the front, which made it a very heavy car – equivalent weight to modern four-wheel-drives.

"We stopped approximately 30 feet before the level crossing at Llangothlin, on the left-hand side of the road, and there were sheep and other animals in the paddock, close to the fence. It was a foggy night, we talked and then dozed off about 3 am.

"I was asleep on the front bench seat and my mate was asleep on the back seat. Next thing I knew the car was going up and down, up, and down, and Michael's being thrown off the rear seat and up in the air, and I'm getting thrown up over the steering wheel!

"And all I can remember is looking out the back window and seeing a big, black shadow over the back tow-ball.

"Anyway, we got bounced around and, lo and behold, the thing stopped, and Michael screamed a few swear words and jumped over to the front, and I hit the motor, dropped the reverse gear, and took off. And as we took off, I looked over my left shoulder and noticed there were no sheep in the paddock anymore – they'd disappeared.

"So, you work it out: how does a guy who weighs 75 kilos, in the back seat of a heavy Valiant, get thrown up like that, hit the ceiling, fall to the floor, up and down again repeatedly, ricochet off the back of the driver's seat? You do the maths, the weights, the physics involved.

"We kept driving non-stop. We honestly could not talk – we just sat there all the way to Tenterfield. And we pulled into a service station there and the guy said, 'What's the matter?' and we were going, 'Th … th … th … that …' [laughs] That's all we could get out!

"We parked under the big trees outside of town and got some sleep, and [later] we just looked at each other and said, 'Nah – we're not going to talk about this one – no one's gonna believe us!'

"That year I did the disused railway survey from Armidale, Guyra, Glenn

> Innes, Tenterfield – right up through there, measuring all the wooden railway bridges, going into a lot of properties along the line.
>
> "Just north of Llangothlin I was walking the line through a culvert and a fox was following me, and I saw his ears prick up and he *bolted*. I kept walking and then noticed a smell like a stinking, rotten carcase.
>
> "OK, I kept walking, but I had a cold sweat on my back; I didn't feel right. So I stopped, made my mark on the rail with spray paint and went back to my car. And then something was telling me, 'Go back! Go back! Go back!'
>
> "So, I went back. I didn't see anything, but later I thought if that smell came from a dead carcass, why did the fox run away – why didn't he go and eat it? It's a free meal."

Later, Neil read of cases where similar foul odours have been attributed to yowies. So he now believes that, given the proximity of the second location to the site of the 1978 incident, he could well have been close to yowies on two occasions.

The very uneasy feeling that he experienced just after noticing the odour, and which caused him to abandon his task, is reminiscent of the "nameless dread" reported by many other people who have been in close proximity to yowies.

Case 204. 1979. Phillip Island, VIC

The witness, Mrs. "M", reported her experience to AYR via email on 29 May 2015.

She told of encountering, in broad daylight, what may have been one of the small, hair covered, manlike creatures that Indigenous people call junjudees. She was 11 years old at the time.

> "It may have been early spring … walking through tea tree near my childhood home, I looked up to see what I thought was another child. I stopped and yelled 'Hi', and put one hand in the air … We did not have a lot of visitors [so] any new company was a thrill.
>
> "I stopped dead in my tracks, just the same as [it] did. In what seemed like a split second it turned and ran the other way up the hill, as fast as I have seen anything run, and he or she was gone.
>
> "I so badly wanted to see and talk to him/her and was very let down at the sudden disappearance, but at the same time I sort of knew that this was no ordinary child and went [to] where I had seen him/her gallop off.
>
> "This child was covered in black hair, was about my size, but was solid in build. I do not remember ever seeing [such a creature] again but do remember

> stories of people who lived among the trees [who she was warned to be careful of].
>
> "The entire island has changed radically … it was all bush then, but the [spot] where I witnessed this incredible sight is between Southport Avenue and Manly Avenue, toward the Esplanade – [it] would be [about] 22 Southport Avenue, Cape Woolamai. I am grateful to be able to share this experience and only wish that I had more."

Undeniably, Phillip Island, situated so close to the densely settled area south-east of Melbourne, and joined to the mainland by an 800-metre-long bridge, seems a highly unlikely habitat for an unclassified manlike animal, so it is interesting to note that that island was the site of one of the very earliest reports in our yowie files – the incident in which a baboon-like animal was supposedly shot. (Case 21)

Case 205. Sept/Oct 1979. 60 km west of Djarra, QLD

The witness, Terry Harding, contacted AYR in early September 2016. He was interviewed by Paul Cropper on 12 September and by Tony Healy two weeks later.

This truly remarkable incident occurred when Terry was working as a 15-year-old jackeroo on "Ardmore", a 2000-square-mile cattle station about 110 km south of Mt Isa.

With a dozen workmates he was mustering cattle at a place called Eight-mile Camp, about 50 km north of the homestead and to the east of the Diamantina Development Road.

Terry Harding

> "This was extremely isolated country, flat plain, a fair bit of gidgee country, quite a bit of timber, but not thickly timbered."

The incident happened "on dusk, just starting to get dark", as they were sitting on their swags, eating their dinner. About 30 feet (10 metres) behind them, the cook's six-month-old Border Collie was chained-up under his van.

Suddenly, the dog began to "scream its head off", they looked over and saw the back of an ape-like creature, "just standing there" looking at the dog, but with his back towards them.

Then "the dog was just lying there, and then, just silence – you could hear a pin drop."

"It just died of fright"

> "This thing loped off, side-on, an ambling sort of trot, but didn't seem to be in a hurry. I didn't actually see its face ... I saw its back and its left-hand side."

The creature was "like a small ape; not that tall ... stooped over – it wasn't standing straight. You know how a chimp or an ape walks, with its knuckles dragging on the ground? If it had been straight, it would have been, probably, five foot.

> "It wasn't skinny as a rake, it was solid, but it wasn't big, like a gorilla, either. There was a lot of hair on it; brown hair – probably a darker brown, like a red kelpie colour. Longish hair, thick – longer than a bear's hair.
>
> "It loped off ... we were right there, but it didn't look at us.
>
> "There were about twelve of us ... there were a few blackfellas ... and some other white ringers, and yeah – we all saw it.
>
> "Everybody was just speechless – we were all looking at each other and our jaws dropped, all goggle-eyed, and like: 'What the bloody hell was *that*?' And the blackfellas were saying 'junjarree! junjarree!'
>
> "We went over to the dog, and it was dead ... I believe it just died of fright: its heart gave out.
>
> "A couple of the blackfellas started following the tracks and then the tracks just stopped – nothing more. I can't say I saw any footprints myself, but I wasn't really looking. [They said] the tracks just stopped – nothing. So it's my belief that they're interdimensional – that they can come into our world and then disappear back into theirs."

The creature was completely silent, and no smell was evident.

Indigenous lore

> "The blackfellas up there call it the *junjaree* – they say there are two different types, a small one and a big one. The big one, that's a plant eater, but the small one, they say is a meat-eater and they're the ones not to get involved with. [Laughs]
>
> "The blacks talked about times they'd seen them before. They said it's very rare for them to show themselves in front of white people – but they see them all the time.
>
> "They said they had red eyes [and that] Aboriginal women say don't go in junjaree country when you have your periods – because that makes the creatures dangerous."

Horses spooked

> "Several weeks, possibly a month, later I had to go back through that area with a mob of horses."

Because the old bloke, Toby Barraclough, who had previously been in charge of the spare horses, suddenly packed up and left, 15-year-old Terry was told that he was now the "horse tailer" – a difficult assignment for a 15-year-old.

So, as the rest of the stockmen walked cattle, two or three thousand head at a time, from one end of the vast property to the other, Terry had to go ahead (taking all the spare horses with him) and set up the next night's camp.

> "I had 50 horses and I had to go back through where that [Eight-mile Camp] was. And I could *feel* it – chills were going up my back and every hair was standing on my head. I think it [the junjaree] was there … it was just … I felt absolutely terrified … I couldn't get out of there quick enough.
>
> "And the horses were reacting – all stirred up, wanting to gallop off in all directions – all real spooky."

A very strange coincidence

Terry went on to mention "something really freaky" that occurred on the very same day as the junjaree incident. That morning, the property owner had driven out to the camp with the mail, which included a letter from Terry's mum, in which she'd enclosed an article from a magazine, probably the *Woman's Weekly*.

As Terry recalls it now (the article having been lost), the story concerned an incident that had occurred in the 1950s at the very spot Terry was working: Eight-mile Camp.

A party of men, surveyors, or road workers, had been camped there, two to a tent.

One night a fellow called Roy, who was half Aboriginal and half Chinese, suddenly screamed, waking up his tent-mate. Something had him by the legs and was dragging him under the side of the tent (which, like most tents back then, had no floor).

His mate grabbed him by the arms and managed to get him back into the tent, but by then the poor guy's legs were very badly gashed.

Min Min lights

The area to the west of Boulia, Djarra and Mt Isa is notorious for sightings of Min Min lights, and Terry is one of the many people who have witnessed the eerie phenomenon – he saw them on a couple of occasions not far from "Ardmore" homestead.

A UFO sighting

When the conversation turned to other strange phenomena, Terry told of a UFO sighting he'd experienced in about 1991 or '92.

The episode began early one evening, "just on dusk", as he was driving out the gate of his family's farm to the north of Murgon, QLD.

The object in question was "a big, round, yellow light coming up in the west." It looked about the size of a full moon, but it was "moving slowly and had revolving lights around the base. I thought, 'Shit – what should I do?'"

Driving to a nearby tableland and looking down, he saw the object again – "I could see the lights of the houses below it, and it was just floating along, hovering above them. It started receding, and then just *took off!*"

Case 206. Summer, 1980. Clyde Mountain, 20 km east of Braidwood, NSW

Tony Healy interviewed the witness, Sheila Eden, on 27 March 2006.

While driving between Braidwood and Batemans Bay, Sheila and her husband began the steep and winding Clyde Mountain descent at about 10 pm.

On about the second or third bend they came upon a large creature that had apparently climbed up from the right-hand (downhill) side of the road and was now crossing it on all fours.

Their first impression was of an enormous dog, but they soon realised, Sheila said, that it was more like "a huge monkey." As they braked and came closer, it turned towards them, raising itself to a "half crouch" and ran up the road past the car. "It was solid, thickset, bigger than my husband. It was not a normal animal."

It was all over very quickly. Sheila doesn't remember seeing a head – just shoulders and long arms that were "hanging, moving, but not swinging." The hair, which shook as it ran, was "browny-black" and about three or four inches long.

The car had almost come to a stop as the creature dashed past the driver's side window, "so close that my husband could have touched it." Even in a half crouch, it was taller than the vehicle: "its shoulder would have been level with the top of the window."

A strange smell lingered for a few moments afterwards, but Sheila wasn't sure if it was related to the creature or to burnt rubber from their tires.

Shaken and suddenly rather scared, the couple hurriedly rolled up the windows and drove on. Sheila is now too frightened to travel that stretch of road by night.

It is interesting to note that in the summer of 1996, just a kilometre further down the mountain, Peter and Belinda Garfoot of Elmore Vale, Newcastle, saw a seven-foot-tall, hair-covered ape-man cross the same road in broad daylight. (*The Yowie*, pp. 57- 58)

Case 207. 1984. Near Springbrook National Park, QLD

Witnesses: Marc Savage, Anna Persi, and Wayne Griffiths.

Marc sent this well-written report (now lightly edited) to AYR on 27 October 2008. Because it concerns what was apparently a juvenile yowie and its mother, the story is particularly interesting.

> "This report is 24 years old, but I thought it would be worth adding to your database of sightings … it took place in the mountains to the east of Wunburra Lookout … near Little Nerang Dam.
>
> "We parked at the lookout, walked downhill and south-east for quite a distance, crossing Little Nerang Creek and continuing into the National Park area. [When] we stopped and picnicked on the hillside overlooking the dam we commented that we felt like we were being watched. Late in the afternoon we began walking back along a very vague animal trail.
>
> "As we were picking our way across the slope of a heavily wooded hill, we crossed a rocky, dry creek bed that ran down amongst the trees. Wayne crossed it first. As Anna crossed, she looked up to her right and froze.
>
> "Wayne turned around and I caught up, and we all saw a human-sized, black, sleek-furred creature. It was squatting on a rock at the edge of the creek bed about six or seven metres away [and] seemed to be playing with a twig or something. It noticed us and I remember seeing a humanlike, yet ape-ish face with glistening eyes. The facial area was black but free of fur, as were the hands and feet. Had it not been moving, it would have been totally camouflaged by the shadows.
>
> "This took only about 30 seconds. The creature didn't seem to be disturbed by our presence at all. Then we heard a strange gurgled sound, and something came crashing through the bush. There was a brief pause and silence while the smaller creature looked behind it. Then a HUGE arm, with a hand and five normal fingers, and part of a body covered in the same black fur, reached out. The smaller creature took the hand, and the arm drew it away.
>
> "We could hear them moving through the leaves for a few seconds, but as soon as the spell was broken, we bolted out of there at world record speed. We had travelled about 100 metres when there was a noise coming from behind us. It was kind of guttural and it rose in pitch each time. It had a threatening tone, and we picked up the pace accordingly!
>
> "Later [they all agreed] that there had been a strange kind of wet-dog smell about the creatures, and that the smaller one was almost definitely a 'child'. It was about 160 centimetres tall, maybe a little taller … wiry and gangly, and its fur was a little patchy, like a chimpanzee's. When it stood up it was surprisingly

upright and erect – nothing like an ape's posture – everything like a human's. Its arms and legs were long.

"There was no real flesh colouring to be seen – it was all black. The bigger creature was the same, but the arm was very muscular. We agreed that we saw a suggestion of a breast as it bent down, so it was possibly a female. We got a split-second look at the bigger creature's face, and we all noticed that it was lighter than [that of the smaller one]. While the small creature's face kind of blended in with its fur, the larger one's was a markedly lighter colour than its fur.

"We thought later that these creatures might have been following, or at least observing us, as we picnicked overlooking the dam."

The Springbrook area is one of the hottest of Australia's yowie hot spots. In *The Yowie* we detailed several other remarkable reports from there, including one involving a dozen witnesses (including a future Australian senator) and another – a face-to-face encounter – involving a veteran National Parks ranger. (*The Yowie*, pp. 2-3, 69-75)

Case 208. Mid to late Nov 1984. Carnarvon Highway, north of Injune, QLD

The witness, Michael McSherry, sent this well-written report to AYR on 26 May 2010. Although 26 years had passed since the event, it had clearly left a lasting impression:

"I am a crew supervisor for an exploration company ... involved mostly in Oz, with some overseas work [so] I have seen a substantial amount of isolated bushland.

"In 1984 I was 23 years old ... working as an explosives pre-loader for a seismic company. That year I witnessed what I always thought to be just stories and myths – a yowie.

"[At around 2 pm] I was driving a diesel land cruiser [towing equipment] which made things pretty slow – about 40 kph.

"As I came around a corner [probably just north of Hutton Creek] and started up a large hill, I noticed [about 300 metres] ahead, on the right-hand side ... a large shape which I first took for a strange looking tree. As I drove closer, the shape suddenly walked out on the road and turned towards me.

"I then started laughing to myself, as I assumed that two jerks were playing games ... I thought that one guy was on the other's shoulders and the two of them were in a ... bear suit.

"This part of my experience still troubles me 26 years later and ... will stay with me till my death. As I approached and the distance slowly shortened, it suddenly swung its torso back around and began to stride off to the other side of the road. When it did this manoeuvre, I was close enough, to my shock and disbelief, to clearly see muscles rippling on its torso and arms, and I knew that what I was looking at was real.

"I will never forget that feeling to my soul and the sensation I felt up my spine. When I think about it even now, I still feel it. Logic was removed from my life that day.

"Its height was 12 to 14 feet, it had very large limbs, with large thigh and arm muscles. The head, though, was not normal to its body size, being smallish and rising to a conical shape on top of its forehead, much different to human heads. Body hair was not real thick, but was definitely apparent, more so on its back at the top. It would easily have been capable of ripping a man's arm right off.

"As it disappeared into the sandy pine scrub it would have been about 75 metres away from me. But it disappeared like a chameleon, as its hair and body colour were the same as the tree bark and stumps. I assume that due to how slow I was travelling it had not heard my vehicle at first.

"I pulled over, got out and began to scan the bushland where it had gone. I thought, 'No one will believe this.' I was excited and scared at the same time. So I began looking for footprints in the sand while looking up and scanning the bush at the same time. I had a strange sensation that it was watching me and was close by. Then I smelt the most horrible stench, like a rotting corpse. Instantly, I remembered watching one of those bigfoot documentaries about the smells that they produce when you are close to them.

"I then realised, as the hair on my neck was [bristling], that I was about 80 metres from my vehicle. I ran as fast as I could, started it, and drove away a changed man.

"I will not camp out alone at night anymore, something I used to enjoy. No one will ever change my thoughts on this matter, I know what is real."

Because the overwhelming fear Michael experienced, which he so vividly described, and which has stayed with him to this day, is so similar to that reported by so many other traumatised eyewitnesses, we have no doubt that his remarkable story is entirely genuine.

Interestingly, Carnarvon National Park, 50 kilometres north of Hutton Creek, has produced numerous reports of *little* hairy ape-men – not yowies, apparently, but rather their smaller cousins – junjudees. (*The Yowie*, pp.127-128)

In July 2018, a Melbourne woman, Mandy Psimaris, experienced a close encounter with a five-foot-tall ape-man on the Carnarvon Highway about 50 kilometres south where Michael encountered the much larger one. (Case 265)

Case 209. Spring, 1985. Near Sunny Corner, between Lithgow and Bathurst, NSW

The witness, Tim (surname withheld), reported this unusually close encounter to researcher Paul Clacher on 9 Oct 2008, and was interviewed by Paul Cropper on 27 October 2008.

It occurred as Tim, then 18 years old, and his brother Jeff were pig-hunting in a scrubby valley on crown land.

> **Tim:** "We were walking abreast, about 12 metres apart. The breeze was blowing onto our faces, and we were trekking slowly, being careful not to make any noise. There were no tall trees, it was just low-lying scrub 3 to 4 foot in height. I was following a track made by the animals, looking down at the valley, and then I looked straight ahead – and there it was: just six foot in front of me!"
>
> **Paul:** "So this thing must've been down below the level of the scrub – lying down, or crouched down – and then kind of popped up?"
>
> **Tim:** "I presume so. It was just ... it was just *there*. I must've surprised him because we were walking so quietly. He had his back turned to me ... I could have touched it, but I was scared ... I didn't raise my rifle ... I had a high-powered rifle [but] I wouldn't have [shot him]. I just looked at him, I thought, 'This is too good to be true!' I couldn't believe what I saw ... I had a very, very, good look at it. I was gobsmacked.
>
> "I'm six foot three. He would have been seven-foot tall and four foot across the shoulders. I weigh 97 kilos, and I estimated him to weigh 300 kilos: he would have been three of me in body mass. It was solid ... like a front-rower. He was [standing up] straight; he wasn't hunchbacked or anything. I saw his arms but not his hands. I didn't see his legs or bum – I just saw [from] his lower back to his shoulders.
>
> "The only thing that was out of proportion was his head [it was] a dome-like shape, like an egg-shape – no bigger than mine. I thought how odd for his head to be so small. And he had no neck; I don't know how he would turn his head to look around – he would have had to probably turn his whole waist and everything.
>
> "His fur was ginger ... like the sunburnt earth, reddish-brown all over – his head, his ears, his back. It would have been about three inches long, but it wasn't

thick and I could see the skin, the pigment … his skin was white like mine, and you know how you get these children or whatever – they're redheaded and they've got those big blotchy freckles? His skin was like that. And his ears – he had funny ears – they came out from the hair. He didn't have earlobes – [they were] just rounded off.

"I didn't smell a thing. [Even though the breeze was coming from that direction.]

"It stood still for probably five seconds. He didn't look at me but knew there was something there … he knew there was a threat there, I guess, and then he just took off, started to move directly away. He ran making grunting noises with every step … you could hear the foot hit the earth and then he would [make a grunt] you know, like when you're trying to get a breath of air?

"He ran manlike – didn't sway side to side – and disappeared into thick scrub. I could have run after it, but I just stood still, overwhelmed by what I saw."

Paul: "Did your brother see it?"

Tim: "No. No – I never told him about it."

Paul: "Really? You must've been busting to tell him!"

Tim: "I was, but my brother is one of those people that would have been looking for it to shoot it, I think."

Paul: "Did you ever tell anybody about it?"

Tim: "No."

Paul: "Wow – you kept it inside for a hell of a long time!"

Tim: "Yeah. I … actually, I spoke to my missus about it yesterday … and she said, 'Did you really see that?' So I sent an email off [to report it] because I wanted to express what I saw and felt. I don't know why, it just feels that this is the right time, sort of."

Tim kept the experience to himself because of his concern for the yowie's welfare and also because of the very profound effect the encounter had upon him. In his initial report to Paul Clacher, he expressed it this way: "I respect what I saw and count myself as one of the select few that have witnessed something so special and so great. All I hope is that he still exists and that man will not challenge or try to invade his world."

It is extremely rare for anyone to get, as Tim did, almost to within arm's reach of a yowie.

Several details in his remarkable story – the almost absurdly small, dome-shaped head set, without benefit of a neck, onto enormously broad shoulders; the creature's immense height and bulk, etc. – conform very well to the testimony of a great many other eyewitnesses.

Another detail he mentioned – that while running the creature grunted with every step it took – has been mentioned in a number of other, fairly obscure, reports since the colonial era, and therefore rings true.

Two other details are more unusual – Tim's is one of just a very small number of reports in which a yowie's skin was said to be pale – and the mention of large, freckle-like blotches is, as far as we know, unique.

Tim impressed us as being an open and honest bloke. Some might find it suspicious that he kept his sighting secret for such a remarkably long time, but he's not the only person to have done that. (e.g., Cases 188 and 213)

Case 210. December 1986. Mt Kembla, NSW

The witness, Ron (surname on file), contacted AYR on 18 May 2006 and was interviewed by Paul Cropper two days later.

Mount Kembla

While engaged in upgrading the Unanderra to Dumbarton railway at the base of Mt Kembla, Ron, a plant operator, arrived on site early, parked in a small clearing and began reading the paper.

> "It was about 6 am, just coming up light, when I had this feeling: 'Somebody's behind me.' I'm a big bloke, but it put the wind up me ... an eerie feeling." Getting

out of the car, he took a few steps and found himself only four feet away from "… a dark shadow – and it was big, probably seven feet tall, broad and strong-looking. The head was like ours, but broader and maybe a bit narrower at the top. I weigh 120 kilos [265 pounds]. This was double my weight, if not triple. It might have been a browny colour; scruffy, hairy – I'd say 100 mil [4 inches] long.

"I froze – I didn't know what to do – run, scream or what. It took off down the hill – hard, rocky terrain, a lot of thick lantana … trees … underscrub … it went crashing through. Long strides. The arms were down near the knees, [they] were swinging. It wasn't a person."

It left "a real horrible smell … like when your dog goes out and rolls in something. Filthy, dirty, musty. I got back in the car and just shook. Even now [20 years later] I'm shaking … goose pimples.

"I was bewildered. All day, blokes were saying, 'What's wrong with you?' and I'd say, 'Nothing, nothing.' I wouldn't tell anybody.

"Then, about two weeks later, about a kilometre up the escarpment, me and another bloke seen it – clear as a bell this time. We were having lunch and my mate looked up and says, 'Jesus Christ – what's that?' Straight across the train line there's a bank that rises up – and this thing was standing there, 50 metres away, just inside the tree line, in clear view.

"I reckon it was the same one – same size and dimensions. It was like an orang-utan – because it was orangey-coloured hair." The head hair was thick, "wild and dirty"; on the chest it seemed sparser, but around the lower abdomen it was so thick that it was impossible to say if the creature was male or female.

"One hand was on a tree, and it was, like, looking at us, saying 'What are *you* doing here?' A face like a monkey … shiny, dark, probably like a gorilla … didn't have a nose like ours … more broad. His eyes were deep, dark … the contact was there between me and him … an aggressive look that was overpowering and scary. I was shaking like an autumn leaf again – waiting for it to come bellerin' down the hill and attack us – that's the feeling you got – [but] it just turned and walked away up the escarpment … big strides, nobody could step that long … torso bent forward slightly. And you could smell that odour as soon as he moved. It's a horrible smell and it stays with you – it's something you don't forget."

Mt Kembla and surrounds have been notorious for yowie activity since at least the 1880s. (e.g., Case 62. See also *The Yowie*, pp. 183, 211, 253 and 289)

Case 211. 1987. Murray River, 14 km north-west of Yarrawonga, VIC

The witness, "S", sent the following (now lightly edited) report to AYR in 2015.

> "Duffy's Beach is on the Victoria side of the Murray .... The banks are pure white sand. It's a wonderful place [for boating or fishing].
>
> "I decided to go there by myself one [winters] night ... arrived at our normal campsite at 11pm ... set up and had a nice big redgum fire going between my 4WD and the river.
>
> "About 45 minutes later I started hearing noises upstream about 50 to 100 metres away ... rustling and branches breaking, and thought it may have been some cattle grazing, bumping into things.
>
> "It got a little more serious when I heard trees being pushed over, and another one, and another one, louder and louder. The fire was big ... so whatever it was, [it] knew I was there.
>
> "I turned the music down and was beginning to get a bit panic stricken. I stood there with a big screwdriver in my right hand and a hammer in my left. Not sure what I was going to do with them but needed something to defend myself with. I had my back to the river, looking towards the bush.
>
> "It paused every now and then, but would have a burst of running ... then another ... going from my left to my right ... pushing redgum trees down ... I could see the tops of the trees falling. I was young and fit [but] couldn't have run that fast even on a flat road. [It was] making groaning noises. What was happening didn't make sense.
>
> "It would stop and grunt ... I could see it through the bush, and it was standing upright. I had a spotlight [in the car] but wasn't game to shine it ... in fear of upsetting it. I'd heard of the yowie before ... in the Blue Mountains, but why would they be out here?
>
> "It came through and stood on the track 50 metres away for a good 15 seconds. It was a lot bigger than me – about 7 foot tall ... massive. I was standing behind the car. I wouldn't move, and it wouldn't move, like a stand-off, and it was looking at me. Then I guess it had had enough and ... ran towards the river ... made a big, almighty splash [and] started swimming towards the other side.
>
> "I grabbed the spotlight and hit it right in the back of the head. The head was a dark colour ... narrower at the top and broad at the bottom. It was doing a dog paddle but wasn't moving its feet. It was mostly under water. The hands would break the surface occasionally, but the head was just gliding along ... swimming

quite easily. I followed it with the spotlight for about 100 metres to the other side, where it crawled up the steep bank and ran off through the bush.

"That night I slept in the car with the windows up. Next morning, I found half a dozen 18-inch footprints on the beach … the strides were about 1.5 metres apart at least.

"I've always been reluctant to share that story. All these years later, I've decided to make contact and share it.

"The only other odd thing I've seen is a black panther down along the Great Ocean Road."

Case 212. 1988. Lost World Plateau, Lamington National Park, QLD

The witness, Tom Clark, contacted AYR on 27 August 2013 and was interviewed by Paul Cropper a couple of days later.

Lost World Plateau, an area of untouched, primal rainforest, is very difficult of access. Tom hiked and camped up there many times over the course of nine years, until he was about 21 years of age.

The incident occurred after Tom, his younger brother and three other boys endured a very hard bushwalk on a "blistering hot day" and reached a small clearing on the very edge of the rainforest. There, despite a big lightning storm and heavy rain, they made camp and managed to get "a pathetic little fire" going.

At about 8.30 pm, as the inky darkness was briefly illuminated by a two consecutive lightning flashes, Tom and his brother were shocked to see a huge, shaggy apparition walking past, right in front of them. It was extremely close: no more than ten feet away, just inside the edge of the forest.

> "It was seven-foot tall, easy [and] pretty solid ... walking like a heavyset person – probably would have weighed 110 to 120 kilos. I saw him from about his thighs upwards and most of his arms. The arms were probably a bit exaggerated in length – but he was a bit hunched over. He was just strolling through the bush; he wasn't too interested in us."

They didn't get a clear look at face: "We saw only it's profile … probably more of an ape-like profile. The hair seemed to be orangey-brown; probably a bit lighter in colour than an orangutan's … like an unkempt English sheepdog – long and a bit matted."

Because of the thunder and rain the boys didn't hear footsteps or notice any vocalisation.

> "We weren't in a position to follow it, and we didn't tell the others until morning because the young ones would have spun right out. In the morning we couldn't find any tracks or anything."

Tom and his brother had heard quite a lot of stories about yowies in the area: "People being followed through the bush and so on. One particular place we used to go to always had a [rank, unpleasant odour] and we used to get the feeling that we were being followed, but this was the only time we actually saw anything."

Lamington National Park is contiguous with Springbrook National Park. Both parks and, in fact, the entire McPherson Range in which they are situated, are notorious for yowie activity. (e.g., Case 207. See also *The Yowie*, pp. 2-3, 67-87, 236 and 244)

As Tom's encounter occurred during a lightning storm, it is worth mentioning that Aborigines from widely-separated regions of the continent – north-west Western Australia, far north Queensland, and the NSW south coast – say that the Hairy Man is seen most often in stormy weather. European pioneers in the "Jingera," below Captains Flat, NSW held a similar belief. (Case 35 and *The Yowie*, pp. 15, 207- 08)

Case 213. Aug 1989. Redbank Plains, QLD

The witness, Michael Serkel, contacted AYR on 7 March 2017. He was interviewed by Paul Cropper on 8 March 2017 and by Tony Healy on 25 July 2018.

On a warm winter's day, Michael and his mate David (surname confidential), then both 14 years old, wagged school and went to a spot they knew very well, on the banks of a small tributary of Goodna Creek, just past the edge of settlement at Redbank Plains.

> "At about 1 to 1:30 pm, after catching some crayfish, we made a tiny little fire. It was a windy day, but we were in a sheltered spot, next to some trees. The area around us was quite open – a few tree stumps, a few bushes: a fairly clear radius.
>
> "We were quite close to some big electrical power lines   the big pylons that cut through the bush.
>
> "There's a little fork in the creek. I was sitting with my back to it, but I felt a bit uneasy for some reason, so I turned [around]. Then we were both looking down at the creek.
>
> "This was the really freaky thing: all of a sudden, the wind stopped and all the insect noise – a constant humming of cicadas – stopped, and everything went *dead silent* – you could literally hear a pin drop. It was an eerie feeling; it almost felt like a weird pressure in the air – like if you walk into a room where two

people have been arguing and you can feel a tension there [and then] there was a noise – a bang.

"Then a kangaroo came jumping out of some bushes. It went past – two hops and it vanished into the thicker part of the bushes. And out of the top right-hand side of my vision I saw a sort of blur … and this yowie – it was quite small – only about the size of what my eight and nine-year-old daughters are now. It looked like it went for the kangaroo. It went past almost in a blur – really quick – I got a side view; it was sort of crouched, a very dark, gingery colour. The arms were really long.

Michael Serkel

"It went past in two blinks [and] noticed us and ducked behind this big cut-off tree that was there – a stump about 80 centimetres high and quite broad."

The stump was on the edge of the little clearing, only about five metres from where the boys were sitting. Crouching behind it, the creature began a weird game of peek-a-boo:

"It had its arms wrapped around the stump and was sitting there looking at us. I could see its hands [which] were really shiny, the fingers were really thick and black, and the face was black too, around the eyes.

"But it didn't raise its face entirely – it was visible only from the top of the head down to about the nostrils. The hair on top of its head was spiked up. It had the freakiest eyes I've ever seen – almost like teddy bear eyes – like buttons, but very shiny. They almost pierced straight through you – it felt like it was hypnotic. I just could not stop staring at it.

"Its face looked monkey-ish in a way, and for a while I told myself, 'This is some monkey that's escaped from a zoo or circus.' But it didn't really look like a monkey.

"I was paralysed with fear: realising what this thing was, but my mind not grasping what I'm looking at.

"I said to Dave, 'Are you seeing what I'm seeing?' And he said, 'Yes!'

"While we stared at it, it would throw glances at us, and then pretend to look down and around – like the gorillas in the Diane Fossey documentaries that I've seen since then. It would hold you, locked in, with a stare and then look down to its left – a deliberate evasion of eye contact. And I was thinking, 'This looks like a baby – not an adult creature.' I didn't think of bigfoot, because at the time I didn't know we had bigfoot in Australia. It wasn't a topic that anyone spoke about back in '89.

"Dave was a small kid. Back then I was small too – maybe 1.7 metres tall. This thing was smaller than me, but it just terrified me – I was too scared to even get up and run – and Dave was the same.

Michael's sketch

"We sat there for, I'd say, almost a minute, but it felt like an eternity – it could have been two minutes or 20 seconds. And [then he and his mate jumped up and ran] I don't think I've ever run harder in my life!"

Strange as it may seem, Michael, like Tim (Case 209) and a few others, didn't tell anyone about the encounter for many years.

"No. The funny thing is, Dave and I haven't spoken about it much either. We couldn't tell anyone at the time because we were skipping school – so we didn't want to tell our parents.

"[The creature] didn't show any aggression ... but just the eyes peering into you was just horrific ... it was almost like it was a supernatural experience – with the way the wind and the insect noise stopped and so on – it felt all surreal and strange. It was a really horrific experience for me."

The small ape-like creature might have been a juvenile yowie, but as it was out, apparently hunting, by itself, it could have been an adult junjudee.

While there are many interesting aspects to this story, one detail that might be quite significant is that it took place very close to some big electrical power lines. While we have not done a statistical analysis, we have noticed that power lines, microwave towers and other electrical installations have featured in quite a few other yowie (and sasquatch, for that matter) reports. (e.g., Cases 232 and 254)

Big cats

Michael mentioned something else that might be relevant – like a surprising number of other people in our files, he has encountered not only the mysterious ape-man, but also a large, cougar-sized cat. He has, in fact, sighted the felid twice. On the first occasion, in full daylight, the creature crossed in front of his car on the Brisbane Valley Highway on the western side of Wivenhoe Dam, near the turnoff to Captain Logan Bay.

The second sighting occurred on the same road, just 200 metres north of the first. On that occasion, Michael's son (who confirmed the story) was also in the car. The cat was within a couple of metres of the car as they drove past, and fully illuminated in their headlights. As it was standing right beside a roadside marker, they could see it was about the size of a German Shepherd.

Later, after consulting various picture libraries, they concluded that the cat closely resembled the American mountain lion, except that, unusually for cougars, this creature's fur was slightly mottled rather than uniformly brown, and it appeared to have rings of darker fur around its tail.

Given that Michael said his encounter with the small ape-man "was almost like a supernatural experience ... surreal and strange", it is well worth mentioning that quite a bit of apparently ghostly activity occurred in and around his family home prior to, and after, that event.

Worth noting, too, is that a similar cessation of insect noise and wind – a sudden, eerie stillness – has been mentioned in many other yowie (and sasquatch) reports over the years.

Case 214. 1989 or '90. About 5 km east of Gympie QLD

The witness, Don Averillo, contacted AYR on 20 November 2015 and was interviewed by Paul Cropper a couple of days later.

> **Don:** "It happened about five kilometres out of town on the Tin Can Bay Road. We lived [about five kilometres further out, on the same road] so we used to hitchhike around a bit. At the time I was about 15, my brother was about 18.
>
> "It was well and truly daylight. We'd walked for about an hour out of town and got to where there was [an intersection – a good place to hitchhike from.]
>
> "It was open country. There were some trees off in the distance, but just scrappy, overgrown paddocks, three-foot-high grass, on either side of the road.
>
> "We were standing there, looking up the road … and this thing just wandered up over the embankment onto the verge and started strolling down the road towards us.
>
> "It was probably about 100 metres away, but you could see it clearly, and we were both just standing there, gobsmacked, sort of thing.
>
> "Then it turned and walked directly across the road, and you could see its little arms swinging, like a humanoid sort of thing. [It looked] really comfortable walking upright.
>
> Then it started walking towards us down the other side and after a while … it got to within about 30 metres, and we heard a car coming – and it just coolly went off the road into the grass.
>
> "I dunno if it was even four-foot tall – probably more like three-to-four foot. It was either very [solidly built] or the fur was very thick … probably an inch or so long. Sort of grey-brown, not real dark, but it was definitely very hairy all over. [They] couldn't make out a face or anything – I can't remember making one out."

Don noticed that the creature's neck was quite short, but he couldn't make out any ears. It made no noise and the lads didn't notice any odour.

> "A lot of people talk about the [big] yowies. Me and my brother have always assumed it was [a juvenile yowie] because we'd never heard of the little ones [junjudees]. "Do you remember that 'Chippy Gobbledok' thing? [A small, hairy, bipedal critter that used to feature in commercials for Smith's Crisps] I always thought it looked like that [laughs]. But my brother – remember that TV show, 'Alf'? He thought it looked like that, but without the big schnozz.
>
> "Me and my brother were brought up in the bush. We used to hunt a bit and

normally we'd have been running up the road to chase it – to find out what the hell it was – but we didn't. The car pulled up and we grabbed the lift and went home.

"And when we got home, we shook our heads, saying like, 'Shit – what the hell just happened?' Because it was bizarre, you know."

Paul: "How long did you have it in view for?"

Don: "Oh, a good couple of minutes, maybe. It was just strolling slowly, so casual – it was unbelievable! [It looked unconcerned] and I remember, I didn't feel concerned either. I think [he and his brother] didn't speak. We might have just looked at each other and, like I say, that's quite bizarre.

"We were sort of spun out by the whole thing. This sounds silly – but it almost felt like it had us in a sort of spell – you know, put us in a sort of trance. But maybe we were just spun out at seeing it.

"We were just absorbed, you know – no room for anything else – just absorbed. We had clarity but were just absorbed, watching this thing – just amazed. But it went a bit beyond that because normally, as I said, we would have walked up closer to check this thing out and look for it – but we just stood there."

Paul: "And you didn't talk about it right away?"

Don: "No. When he disappeared into the grass the car pulled up and we jumped right in – and that was part of the bizarreness – because, like I say, we were hunters as kids and normally we would have been up there after it, so like I say, it was almost like it was a bit of a mystic spell – and the fog came off when we got home. Then it was, like, 'Holy hell!' But then it was too late.

"Me and my brother still talk about it, like: 'Remember that thing that we seen?' But what can you do? You've seen it and that's all there is to it."

Strange daze

Don's feeling that the little creature "had us in a sort of spell: you know – put us in a sort of trance … a bit of a mystic spell" is very interesting in view of what Michael Serkel (Case 213) said about his Redbank Plains junjudee encounter – that it "was almost like a supernatural experience … surreal and strange."

Case 215. c.1990. Near Jenolan Caves, NSW

The witness, Luke Simon, contacted AYR on 6 September 2006 and was interviewed by Paul Cropper a week later.

McKeons Creek runs through a gorge just outside Kanangra Boyd National Park.

Twenty- six-year-old Luke was quite familiar with the area. At 10am on the day in question, he "decided to go hunting for a couple of rabbits ... walked down ... and saw a movement out of the corner of my eye, downhill, to my right, a good hundred metres away. It's native bush, a lot of fallen trees, difficult to walk through, but I did have a clear line of sight.

> "At first, I thought it was a deer, because it was big, but when I took a closer look ... I thought, 'That's not right!' It was humanoid ... like a giant man in an ape suit. I didn't see its face, only its back as it was going off through the scrub. And I brought the rifle up – just instinct – and actually took a shot at it. But that didn't have any effect whatsoever – it just kept going. I was in such a state ... open sights ... I probably missed.
>
> "I'm six foot two; this was at least two metres [six feet seven inches] tall – like nothing I've seen before in my life – not even a gorilla. I'm 110 kilos. If I could make a stab, [it would have weighed] as much as a horse. It was huge – it could easily take me apart.
>
> "From shoulder to shoulder it would have been four feet, and the head ... was very big. Quite long hair [all over the body] like a matted dog. Dark brown. I saw its arms: long, very thick, about the same [relative length] as a human's ... didn't see any swinging motion. The only thing that struck me as being different was its neck: it didn't appear to *have* a neck, as such. It walked like a man.
>
> "The *smell* of it – that's what struck me almost immediately, was the stench – it stunk really badly. Like a dead animal, very strong, pungent."

Thoroughly rattled, Luke hurried back to his vehicle and drove away.

Case 216. 1991. Australind, WA

The witness, Michelle Netos, contacted AYR on 16 November 2011 and was interviewed by Paul Cropper five days later.

At the time of the incident, she was just under 16 years of age.

> "[Collie River] runs through Australind, through a lot of paddocks and bushland. I used to go down there swimming almost every day through high school. There are a lot of wild berries growing there, also wild pigs, etc. We used to go hunting down there. I'd been down there two or three times previously at night – but always with someone else.
>
> "So, one nice [moonlit] night, I thought I'd go for a walk down there. It's about a 30-minute walk through bushland, on dirt tracks.

"I was sitting down at the river, and normally it's very peaceful, but that night there was something ... it was like everything went quiet ... I had the sense that I was being watched and I just didn't feel comfortable, and it got so strong. The fear started ... almost like something was saying, 'You've got to get out now – or you're not getting out!' [Laughs] An extremely strong vibe – it built up pretty quick. I've never felt anything like it before or since.

"Between the river and the trees there's nowhere to hide – it's all grass – maybe 200 metres. So, I got up and [walked across the open area to the start of the bush track.]"

Through the trees, the track was sandy and clear of sticks, so she made very little noise as she walked.

"Then I heard [from the bush to one side], *crunch, crunch* – like someone walking across sticks. I stopped and looked to my right and couldn't see anything. So, I walked again, but cautiously, and again I heard *crunch, crunch, crunch.*

"I stopped again and then I could see an outline of something ... well over seven feet tall – even closer to 8 feet. My boyfriend at the time was six foot five and about 80 kilos. This thing would have *towered* over him and was a lot more solid. I was probably only about seven metres away from it – so that's why I could see so much of the shape and the size ... I couldn't see its face though.

"It looked like a man, not that I could see it well – it was more like a shadow – I only had the moonlight [but] I could definitely make out hair – two or three inches long and not just on its arms – it whole face and head seemed to be covered.

"It was standing still, like it thought it was hidden in the trees and shadows, but it was a bright enough night for me to see the shape. And I remember thinking, '*What* am I looking at?'

"It wasn't human. The closest thing I can compare it to is the Wookie, Chewbacca, from 'Star Wars'. Except Chewbacca didn't have the shape this had – [this] had more muscle tone. It was like a human shape but more muscular – but not like a bodybuilder; solid but lean, like a fit athlete.

"I took a few more steps to see if I was seeing things. And as I moved, it walked: *crunch, crunch, crunch.* And when I stopped, he stopped. And so, I thought, 'Is this thing *hunting* me?'

"I didn't want to run. I thought, 'If I run, this thing is gonna get me'. So I walked as quickly as I could, trying to keep as calm as I could. As I walked, I didn't take my eyes off it [and] I could clearly see it moving.

"But the track ahead bent to the right, and I thought, 'If I go right, I'm going to meet this thing – it's going to come out onto the track!' So I started running – I thought, 'I've got nothing to lose.'

"I finally ran up to where I'd left my bike. I'd put it in the bushes, so no one could see it. But the really weird thing was, the chain had been taken off and laid down on the grass and there were sticks poked into the gears and where the chain goes, as if something had been poking at it to see what it was. I'm wondering [now] if it had watched me the whole time.

"I just grabbed my bike and ran to the road – only about 10 metres to the road – and then just took off to the nearest streetlight, the nearest house. I never went to that site again. Even if I go back to Australind to visit my family, I won't go near it.

"Until I saw [the AYR] website I'd never heard of people having similar experiences. I knew there were Aboriginal stories about yowies, but I thought they were just legends or something spiritual.

"Now I wish I'd gone back and looked for tracks. I think if it was a yowie, I don't think it wanted to hurt me – because it had every chance to. I think it was probably as curious of me as I was of it. I told my family about it and I don't think they disbelieved to me, but they didn't encourage me to talk about it either.

"The whole thing probably lasted about 20 minutes – 20 minutes of thinking you're being hunted by something.

"It stood like a human, it walked like a human and it ran like a human – but it was not human. There's no way this was a human. It was something I'd never seen before in my life. It's stayed with me all these years."

In addition to the creature's general appearance, there are several things in Michelle's report that suggest she really did encounter a yowie – details that have cropped up time and time again in eyewitness reports from other parts of Australia.

These include her recollection that "everything went quiet" followed by a sensation of being watched, which morphed into an overwhelming terror that was "almost like someone was saying, 'You've got to get out now – or you're not getting out!'" The way the creature walked abreast of her, just off the track, also tallies with the behaviour of yowies in many eastern Australian locations.

It may be worth noting that the Collie River is only 25 kilometres north of Donnybrook, where a yowie was sighted in 2011. (Case 245)

Case 217. 1992 to 2022. Near Smithton, Tasmania

Dean Harris contacted AYR via Facebook, on 29 Oct 2018 and was interviewed by Paul Cropper three days later.

He told of repeated interaction with yowies over the course of 30 years, almost always in a large area of bush, cut through with streams and gullies, to the south-east of Smithton.

The first incident occurred in 1992, when he was 15 years old, spotlight shooting near Alcomie.

> **Dean:** "We always hunted there as kids, and always felt there was something [strange] there. The dogs would always be looking back and were a bit flighty. [On this particular night] there were three or four of us, and we saw a set of eyes about 60 metres away. We knew the eye shine of all the normal animals, but these were really strange – they looked like they were about two inches in diameter and four inches part, and reddish. They moved around a bit and we could see the outline of [a big upper body].
>
> "We sat there looking at it through our scopes for a good 20 minutes, and as soon as we started [driving] towards it, it just bolted off. I was standing on the back of the truck, and when we got over there [he realised that] I would have been eye-height with it.
>
> "[About six months later, in the same area] we heard these footsteps going boomp, boomp, boomp, boomp, over the bank, and we ran down to look, and all we could see was a black figure … the back of the head and shoulders. It looked fairly tall. The strides of the running sounded wide and heavy. It *moved* – it was fast.
>
> "There was nothing for about 12 months, and then were out lobstering and we came across two footprints – like it had stepped down off the bank into the soft stuff, tried to step over the creek and stepped back into the soft stuff again. It was a five or six-foot stride. Nobody would have been walking in the bush there in bare feet. And these feet were pretty bloody big.
>
> "About ten minutes later we heard this noise in the bush. We knew all the animal noises, and it wasn't any of that. Then we were out skinning wallabies, and we heard the same noise again. That's a noise you don't forget. Sometimes it's so loud that it vibrates in your body – sounds like a cross between a freight train horn, a bellow of a bull and a roar – hard to describe – you can hear it from a long way.
>
> "Recently, we were at place out there where nobody had been. There were no motorbike tracks, and nothing had been moved. We found more footprints of about four different sizes in the leafy undergrowth. On one, just the heel would

> have been four-and-a-half to five inches wide. I lay down my nine-inch knife, and that was lucky to go halfway along the length of the foot.
>
> "Some of the prints were ten to ten-and-a half inches long, some 16 to 17. But the 'big fella' – his tracks were 19 to 23 inches, easy. *Huge!* Some tracks were two to two-and-a-half inches deep, some shallower. In the same area we have found stick formations, tee-pee shaped structures and bent-over trees."

In 2018, in the same area as the 1992 incidents, Dean and his mate Wayne found several trees – large man-ferns – "stripped of their leaves. One was broken in half and had scratch marks all the way down. The leaves were scattered around, and it looked like something large had laid down there."

Dean and Wayne then found a tee-pee structure and a big arch made from trees bent and woven through other trees. As in some mainland yowie hot spots, they have had stones thrown at them – or at least in their direction – causing them to rapidly retreat.

Dean has glimpsed the creatures on a few occasions, but his best sighting occurred as something followed him and a mate though thick bush, always just behind them and to the left. After they broke into a run and Dean's mate dropped behind, hoping to see the creature, it skirted around, and Dean found himself face-to-face with it in a small clearing. As it paused, just four metres away, he got a good look at it from head to toe. "Full frontal – the whole body."

Although it was seven or eight feet tall, he had the impression it "wasn't a full adult." It was completely covered in black hair, "blacker than charcoal," and its face, where nor covered with long, coarse hair, was also black, and shiny. Its large eyes were "absolutely jet-black, with no white in them – completely black." The nose was broad, there was "not much neck" and the arms didn't seem disproportionately long.

Although the huge creature didn't look threatening, Dean "was frozen." Then his mate called out to him – and things began to get really weird: "I tried to answer, but ... I know this sounds crazy ... I think they 'mind-speak' in your head, like broken English – it was like he [the creature] was saying, 'Don't answer!'" As Dean stood there, "frozen", the yowie rushed away.

> "I felt like I was going to faint, and when [his mate] arrived, he said I was all pale. I felt like I'd been hit by a truck."

While we have documented many reports of ape-men inducing abject terror or brain-fogging confusion in witnesses, the only other mention of apparent "mind-speak" in this collection is in the testimony of Neil Watterson. (Case 203)

That utterly strange and unsettling phenomenon has, however, been reported on several occasions by North American sasquatch witnesses (e.g., Thom Powell, *The Locals*, pp.136-37), by Russian *almas* researchers, and once, to our knowledge, by a researcher in Nepal. (*The Yowie*, p. 185)

Apparent "mind-speak" isn't the only "high strangeness" Dean mentions. At night, he says, the creatures' eyes "self-luminate – they shine even if there's no moon, no light shining at them. They're usually off-dull yellowy, but when they get agitated, they go orangey-red – it's more than weird, it's unnatural, like something out of a movie."

Perhaps the most notable thing about Dean's experiences, however, is that they occurred in the island state at all. Strangely – or perhaps not so strangely – we have only four other Tasmanian reports in our files, and virtually no Tasmanian Aboriginal yowie lore. (See our comments accompanying Case 199)

Case 218. 1993 or '94. Brown Mountain, 15 km south of Nimmitabel, NSW

After noticing an article about the incident in a regional paper – *The Magnet* (Eden), 6 May 2017 – Paul Cropper contacted the witness, Mark Dowton, and interviewed him.

Late one moonlit night, as Mark, his brother, and a friend were driving to Eden on the Snowy Mountains Highway, their car swerved off the road near the top of Brown Mountain and slid down a bank. As they were unable to extract the vehicle, Mark's brother hitch-hiked to Eden to arrange a tow, leaving the others to sleep overnight at the cold, remote spot.

> "It was a bright, moonlit night," Mark recalled, "with birds and frogs squeaking away in the scrub." Then, between 11 pm and midnight, as he was half-dozing in the front seat and his mate David asleep in the back, all animal noises suddenly ceased.
>
> "That made me look up – and then I went, 'Oh, my God!'"

A powerful-looking, bipedal figure, covered in black hair "like a gorilla", had approached to within about fifteen feet of the vehicle.

> "He was a good five-and-a-half foot tall; very well built … fairly wide across the shoulders … his arms were probably a tad longer than ours. Its face … sort of black skin … was like a gorilla's. It was *not* a human.
>
> "He was standing still, not moving his feet, but swaying from side to side. He just kept staring at me. Mate – it was the scariest thing of my life! Snakes, spiders, any animal doesn't bother me, day or night, but this thing, it petrified

me – I couldn't move! We had rifles in the car, but all I wanted to do was lean over to the back seat and tap David and say, 'Quick – have a look at this!' But I couldn't speak, I couldn't move."

Without making a sound or exuding any noticeable smell, the creature continued staring at Mark for what seemed like "probably 15 minutes or more – a long time – then it *slowly* turned around and walked away – real casually.

> "We stayed there the rest of the night but nothing else happened, and in the morning we looked around for tracks, but it was all grass and bush, and we couldn't find any signs.
>
> "This was no joke; this was real. There was no alcohol [involved] nothing like that. I'll never, ever, forget it."

Mark's sighting seems to have occurred within a couple of hundred metres of the spot where part-time ranger Chris McKechnie almost ran over a seven-foot yowie in 1997. (*The Yowie*, pp. 66-67).

Also, as detailed in Case 130, George Summerell, of Nimmitabel, rode up close to a huge, hairy ape-man near Packers Swamp, just 10 km south of Brown Mountain, in October 1912.

Case 219. 1994 or '95. Bonalbo, NSW

Kris Cook, an Aboriginal man now living in Ballina, contacted AYR on 21 May 2017 and was interviewed by Paul Cropper a few days later.

Twenty-three years earlier, he and his brother had seen a yowie in the village of Bonalbo:

> "Because our parents had to go to work early, my brother Josh and me had to [wait near the school until it opened]. I was about 10 or 11 and Josh was six or seven. It was pretty foggy, a winter's morning, and we were sitting there on an old gum tree stump, and we heard this really loud, thumping sound.
>
> "I looked around, and it got louder, and we both looked up at the dam [which provided the town's domestic water.] We were probably one hundred and something metres from the dam. And looking up, we saw this massive, big figure running across the top of the dam wall.
>
> "Because of the way the sun was coming down through the fog, it was like a silhouette, but what we saw was an eight-foot-tall, very strongly built, very solid figure, with 4 to 5 inches of hair streaming off its back as it ran. *Huge!* And running really quick.

"The running sound was like a clapping. The legs were the same thickness from the thigh right down to the ankles, and it had very long arms, and it was running faster than any human being could. We didn't see its face – just a side-on view.

"We had it in view for a good minute or so, because it stopped at the end of the wall and started going up the hill. It was just walking by then and then it just disappeared in the fog.

"We just looked at each other and said, 'We've just seen a Hairy Man!' Because, being Aboriginal, we knew what it was. Living out there you hear all the stories about the Hairy Man … stories from Woodenbong and all through the ranges there … and to see one was absolutely amazing! And we just ran home and told mum and dad – they were just about to leave for work."

As Kris mentioned, the region where he grew up is quite notorious for Hairy Man activity. In November 1976 and August 1977, for instance, yowies were seen in the back yards of two houses in the village of Woodenbong, about 45 kilometres north of Bonalbo. (*The Yowie*, pp. 236 and 239-240.)

Case 220. 1995. Howard, near Burrum Heads, QLD

Mark Smith contacted AYR on 24 November 2016 and was interviewed by Tony Healy two days later.

The particularly interesting encounter occurred as Mark was driving from Torbanlea to Howard on Burgowan Road.

"The weather was okay; it was just dark enough to have my lights on – about six o'clock. [The experience] has stuck in my mind for all these years … it's as vivid as …

"I take it quite slow through there because there's a few wallabies around. My ute had good lights and I was probably doing less than 60 Ks when I saw, out of the corner of my eye, something on the right, and backed off my speed.

"It was a dingo, its ears were back, and it was *going* – full pelt. I had my windows down, and, as it cut across the road, I heard this howling, howling, *howling*. Then this half-grown dingo – more than a pup – [emerged] only five or six metres behind the adult. And it was just *screaming*, and it was *moving* too!

"By this stage I'd come almost to a rolling stop and I'm thinking 'What's wrong with these dogs? They're apex predators.'

"Next minute, here's this thing … it was massive, it was *huge*. It seemed to be down … not on all fours, but hunched over, like it was using its front legs to help it run.

"As it came across behind this pup [and got to within] maybe four feet of it, it made a swipe at its hind quarters with its right hand. And this pup just spun, rotated in the air, and in one fluid motion this thing just grabbed it by the neck, stood up on its back legs, kept its stride going, and just ran away. The pup was just dangling in its hand – dead, I reckon – he must have snapped its neck when he grabbed it. Within two or three seconds it had just blended straight into the left-hand side of the road [where the railway reserve is].

"This happened, all lit up, less than 10 metres in front of my ute [but because the creature crossed at an angle of about 45 degrees] I never got a look at its face – just saw it side-on, hind-on.

"It was a dark, chocolatey-brown. I was in a small 4WD ute, and when it was hunched over, crouched down, its back was almost at my eye level. On its hind legs, I'd estimate it was *at least* seven, seven-and-a-half foot. Big in the shoulders, but quite lean in the lower end of it – quite athletic.

"It had massive, long, back legs; the stride was amazing. You know these athletes – these runners – the six to seven-foot-tall athletes? This thing was like that – in just seconds it was gone.

"It's not the first [yowie event] in the area. I've heard they eat meat, but I was surprised to see it hunting dingoes. You'd think they might live in harmony with each other – but obviously not!" [Laughs]

Mark's sighting occurred about one to one-and-a-half kilometres short of the Brunswick Heads turnoff. The dingoes and the yowie crossed the road from his right (the north, or north-eastern side)

Although there was (and still is) a considerable amount of bush to the north of the road, the animals crossed at a place where the road is bounded on that side by open grazing paddocks.

Those paddocks are separated from the road by a standard three-strand barbed-wire fence, so Mark assumes the dingoes scrambled under the bottom wire and the yowie vaulted the top.

Other yowie events

Mark's encounter happened only eight kilometres south-east of where the Erickson family (Case 235) experienced their remarkable 2007 sightings. Mark heard about those events at the time.

> "Yeah, I know that area, and I once went looking for a [new client's] house down there but got lost and went quite a few Ks into the bush. I got out of my ute and went for a bit of a walk, and found an area about 30 foot around that had been absolutely *flattened*. Eight-to-nine-foot saplings, all just pushed down. And I thought: 'I'd better get out of here – something's not right!'"

Rock-throwing, foul odours

Mark had other interesting experiences while living on a bush block on Burrum Heads Road, about three or four kilometres from the site of the yowie/dingo event:

> "One night a bloke, Ray, who was building a house about a kilometre away, came up and knocked on the door. He was panicking and he says, 'Oh mate, someone was throwing rocks at my window. and I went out and couldn't see anything, but could hear something in the bushes.'
>
> "One afternoon I smelt [something like] burnt Bakelite, almost a burning, urine-like smell. My bull terriers went absolutely ballistic, and we could hear this crashing through the scrub; I reckon there was more than one of them.
>
> "The next day I mentioned it to a work mate who had a property out at Dundowran [about 12-15 km to the east]. And he said he'd had the same experience earlier that evening, about 4 pm – loud crashing, a smell that was 'thick in the air.' His wolfhound took off after it, and he thought he'd never see it again, but it eventually came back about three in the morning."

It's very interesting that Mark compared the smell to that of burning Bakelite, because several other yowie witnesses have mentioned smells reminiscent of burnt electrical wiring – and one (see *Out of the Shadows*, pp. 128-29 and *The Yowie*, p. 144) specifically mentioned burning Bakelite.

It's also interesting that Mark thinks he may have been close to yowies on two other occasions when he was much younger. Once, while he and some other boys – all about 13 to 14 years old – were camped in Numinbah Valley, they were frightened by "... quite a lot of noise and stuff that was hitting the tent. There was howling and screaming and heavy, hard breathing.

> "We all knew what it was. Even the old timer who owned the property said [later] 'Yeah – well, that's why you don't go into the bush at night.'"

Numinbah Valley, in the Queensland/NSW Border Ranges, has long been notorious for yowie activity.

"Another time, when I was in my early 20s, hunting near Bonshaw [further west on the QLD/NSW border] we had a real eventful night – woken about 2.30 in the morning – the [vocalisations] shook us through the tent. Old-timers said there was something there."

Case 221. 1995. Roxby Downs, SA

The witness, Jarrod Nicholson, sent the following (now lightly edited) account to AYR on 17 August 2006:

When he was 11 years old, he, his mate Fred and three younger boys rode their bikes to a secluded dam in scrubby, semi-desert country about 30 minutes from Roxby Downs.

> "We were there no longer than 20 minutes when the youngest member of the group complained that someone in a fur suit was stalking the bush around the huge clay pan we were in … but nobody else saw it. About an hour passed … before me and Fred saw it running … crouching … to another hiding spot. [It] stopped behind a small dune and kept watching.
>
> "The three younger lads started heading back [but] Fred and I were too curious to leave and crouched near the dam … watching him. Ten minutes passed with no movement, so we proceeded carefully towards the thing. We … closed half the gap, when the beast became enraged … and stood up, revealing its true size.
>
> "The thing was around six, maybe seven foot, very lanky and skinny, with tannish coloured hair or fur similar to a camel's covering its entire body except around the eyes. I was too scared to get any better … look at it.
>
> "He started towards us and that was it. We turned and ran as fast as we could to the bikes … pedalling our hearts out, not looking back for a second. Halfway home my chain fell off … I screamed for Fred to stop, but he kept going ('Better you than me', I suppose). I just dropped the bike and ran all the way home … I was in trouble with the old man for leaving my bike behind.
>
> "That was … the scariest thing to ever happen to me (and I've seen some scary things working underground for the last four years). At the time we believed that it wanted to attack us, but now I think he was just after a drink and us little bastards were in his way."

In *The History of Australian Exploration From 1788-1888*, Ernest Favenc wrote about two squatters named Oakden and Hulkes who, while searching for good grazing land to the west of Lake Torrens in 1851, were told by Aborigines that ape-like creatures were sometimes encountered in the area (Case 23). Roxby Downs is 80 km west of Lake Torrens.

Case 222. c. 1996. Barron Falls, on the outskirts of Cairns, QLD

This report, though second-hand, is so interesting, and came from came from such a well-qualified (and quite evidently sincere) informant, that we didn't hesitate to include it in this collection.

When the informant contacted AYR, he fully identified himself but because of his occupation (he is a paramedic) asked that his name be kept confidential.

The following account was compiled from correspondence between the informant, Dean Harrison of AYR, and Paul Cropper between 17 and 22 May 2006.

> "About ten years ago [on a fine, sunny afternoon] we were called to a distressed 50-year-old female who was four kilometres along the Barron River Falls/Power Station walking track. We reached the location on foot."
>
> [The track traverses a thickly wooded area – largely tropical rainforest.]
>
> "We located the female, who was in hysterics – crying, shaking uncontrollably and in shock. She was ... as white as a ghost and could not move. I telephoned her daughter who [soon] arrived. We stayed on the track for about 30 minutes comforting the female. I learned from her daughter that she did not suffer from any illnesses or mental health problems. She lived in [nearby] Caravonica, was an avid walker [and] had walked the track every day for the past five years. We managed to get her out and back to the car park [where] she settled down and was able to speak.
>
> "She stated that she was walking along the track when she observed a large 'creature' walk from right to left, across the track to the opposite bushland. She described it as very tall, with thick, matted, dark brown hair ... its face was like 'dark leather', but she couldn't [make out] any features.
>
> "She stated that the creature made a lot of noise breaking branches as it walked downhill ... she froze ... had a 'panic attack,' collapsed, and was unable to breathe. She had never seen anything like it in her life, and said it wasn't any animal from the Australian wilderness. She was found by another walker, who telephoned us.

Barron River Gorge

"I do not doubt this female's encounter, and I still think about what she told us. I cannot describe in writing the fear I saw on her face that day. I was a sceptic, but I believe she had some sort of encounter with a 'bigfoot' creature. She was transported to the hospital, and I don't know what happened to her afterwards.

"I spoke with Queensland Police in the area, who [said] they had heard of several encounters of a similar nature. One officer told me that a lot of howls and unusual grunts had been heard from the rainforest."

Case 223. Late 1990s. Cedar Grove, Keiraville, NSW

The witness, Petra Van de Moosdyk, reported this remarkable episode to AYR on 5 April 2018 and was interviewed by Paul Cropper a few days later.

The events occurred while Petra was house-sitting a residence in Cedar Grove, a cul-de-sac at the foot of forest-covered Mt Keira. That mountain, along with neighbouring Mt Kembla, has featured in several yowie reports over the past 140 years or so. (e.g., Cases 210 and *The Yowie*, pp. 183, 211, 253 and 289)

Dense forest extends right down to the back fence of the block in question, as it does with many of the neighbouring properties. There is a ravine quite close by.

The residence belonged to an elderly couple. The husband, a retired vet, mentioned something strange before handing over the keys – "That they'd had 'visitors' coming through the yard – he called them 'gorilla people' – and that's about it. I wasn't shocked or frightened when he said that, I just sort of [accepted] it.

"He just said to keep the doors locked [including] the garage door, otherwise they might come in and snoop around – just to be careful. But I actually didn't expect to see anything – so I was surprised and shocked when I did."

The first sighting

"It was in the middle of the day … I went out to check on their goat, [tethered in the vacant block next door]. I was talking to the goat and feeding it, and I heard voices, like a whistle or singing or something, and turned around and here was this young person – this creature – all covered in white fur.

"[The bush] was quite thick but there was an area of very thin trees, all very close together, and it had come down through there. I think it was heading to the fruit tree to get something to eat. It was pretty much halfway down the hill when it looked up and spotted me and sort of went, 'Eeeee!' It did a U-turn and went off, under cover, quite quick. I watched it go up through the trees but in no time he or she had disappeared. I didn't follow it."

Paul: "How close were you to this thing?"

Petra: "Oh – probably about 10 feet. [It was] about five feet tall, maybe a bit more … equivalent to a teenager in build, in height, and probably in age as well. It wasn't muscly [but] not thin either. The hair was white, fairly long, quite shaggy. It wasn't pristine white … pretty much as if you had a long-haired white dog that needed a good bath.

"Its face was all covered in hair. I only noticed the eyes, and to me they seemed like they were going vertical instead of horizontal, you know, like opening up. It had just a short neck. I didn't notice any ears."

She didn't detect any odour, and, because she didn't notice any male genitalia, she got the impression the creature was female.

Second sighting

"I was driving home [in the early afternoon] … crossed the cul-de-sac, slowed down, and was about to hit the driveway when I saw the [same creature] picking fruit and eating it. The tree was just off the road.

"He saw me coming and jumped down into a concrete gutter about 18 inches deep, [near] the fence … and I drove in and parked and walked back down – and it was still there. I could see its feet and its calves and, well, the whole back. I couldn't see its face because it was lying down [on its belly] hiding its face.

"It was lying very still [and] I thought, 'Should I reach down and touch it?' But I didn't. I was very quiet, so it must have thought I wasn't there anymore, and it jumped up and ran back up the hill."

Again, she didn't notice any odour.

Third sighting

"It was in the morning; I was at the kitchen sink. I could see [through glass doors] onto the patio and up the hill. In the backyard there was a table. And I saw these two black creatures come through a gap in the fence, having a look to see if anyone was around.

"They were about 50 feet away. One was very short and all I could see was just its shoulders and head. It could have been five feet tall or a bit under, but then again, it might have been crouching down.

"The other was a lot taller, and I could see from [about] the waist up. [It was] probably the height of an average man: five foot six to five foot eight. They were both covered in black hair.

"I saw the face of the short one, and it was like it had big eyes – like if you open your eyes quite wide – and the mouth looked more like a chimpanzee's – you could say it looked quite a bit like an ape.

"[It] didn't have much of a neck but the tall one had, like, an average human neck. The short one was [thick-set and] would have been quite heavy [but] the tall one [might have been] about the weight of a well-muscled man. I didn't see any male organs or anything.

"They walked upright. The short one sort of waddled like an overweight woman. I imagine it possibly was the female. The tall one walked normally.

"They stayed near the table. [Then] saw me in the kitchen and decided they weren't going to go any further and went back – and that was the last I saw of them."

That episode lasted "Only two, three minutes, something like that."

Paul: "Did you tell the people who owned the place what you'd seen?"

Petra: "No, I don't think so, I don't remember. I haven't mentioned it much. I told my mother and mentioned it once to my sister-in-law, but it's not something I talk about freely."

Paul: "So it made quite an impression on you."

Petra: "Yeah, yeah. I'm only glad I can speak to someone who knows what I'm talking about. There's a lot of sceptics out there and you don't want to talk to people who don't believe ..."

While Petra struck us as being an honest person, and while the site of her experiences is right on the edge of a vast mountainous area that has been notorious for yowie activity since at least the early 1870s, her story is very curious in some respects.

It seems strange, for instance, that the property owner, a qualified vet, would have had such a casual attitude to repeated visits by "gorilla people" – creatures clearly unknown to science. He must have known that, if he'd kept a camera handy and taken some clear footage, the story would have "set the world on fire".

Petra, too, seemed oddly unexcited by her experiences – somewhat indifferent to the importance of what she'd witnessed.

Interestingly, though, the strangest of all the strange things she mentioned – that the young, white-haired creature's eyes "... seemed like they were going vertical instead of horizontal, you know, like opening up [sideways, like a reptile's]" has been mentioned once before. (Case 259)

Case 224. 1997. Picnic Point, Toowoomba, QLD

The witness, Rachael (surname on file) contacted AYR on 12 December 2014 and was interviewed by Paul Cropper on the following day.

The incident occurred on a sunny afternoon, as 37-year-old Rachel was walking, "feeling pretty relaxed", on a track to the Picnic Point lookout, on the outskirts of Toowoomba:

> "It was walking through the long grass below me on the hillside – about 15 to 20 feet away. He was striding along – pretty long steps – with his arms swinging by his side … around seven feet tall; black fur as on a gorilla. Like a man with long arms – arms to mid-thigh. I didn't get much of a look at its face.
>
> "I sort of glanced at it sideways, walked a few more paces and then thought, 'What's that?' And by the time I'd stopped and looked around he'd sat down [with his back to her].
>
> "I felt a bit frightened and felt like I was going mad, really. I thought, 'You can't be seeing this!' It didn't look aggressive; it must have known I was there before I saw it, but I got the sense that it didn't mean me any harm.
>
> "I stared at it for maybe half a minute, then started making noises, trying to get him to turn around and nothing happened, and then I thought 'Maybe that's not such a good idea.' And I got scared and left! [Laughs] He wasn't going to budge anyway."
>
> "I walked up to the lookout atop Picnic Point. Once there, I found I couldn't see its location from above. Many people were also looking across the valley, at hang gliders over Tabletop Mountain. I think the yowie was also watching them."
>
> **Paul:** "What did it resemble most?"
>
> **Rachel:** "Well … nothing really … but like a gorilla with long arms and long legs. The face wasn't black like a gorilla's – maybe like a tanned human face."

Having seen the face side-on for only a few moments, she couldn't say whether it was more ape-like than human-like; she didn't notice any ears.

It walked like a human and "was very muscular … very broad, big shoulders. I'd say a similar weight to a gorilla – but taller, so he might have weighed more. Definitely not human."

The hair was "dense, like a gorilla's fur – reasonably long, but not flowing in the breeze or anything like that."

It sat with "a very straight back. And it was sitting still – rock still."

Strangely, it seemed to make no noise when it walked. "No – that was another weird thing about it: I couldn't hear anything at all."

She didn't talk to anyone else about what she'd seen, "except, very vaguely, to someone at work, because he was interested in the subject too – so I didn't feel too stupid."

Case 225. c. Feb 2000. Yowrie, NSW

The witness, Lynette Gray, contacted AYR on 15 Sept 2010, and was interviewed by Tony Healy a few days later.

The incident occurred in the middle of a clear, sunny afternoon, as she was returning from a bush walk to her property, on a dead-end road a couple of kilometres south of Yowrie.

> "I had just got back onto a wombat trail I had started from and was looking at the ground. And as I lifted my gaze … coming towards me was the yowie … perhaps 30 metres away. It was also looking down … I thought it was possibly smelling my scent or seeing my prints.
>
> "I saw it before it saw me … it lifted its head and quickly ran towards a tree.
>
> "[It was] approximately six feet in height, its face was ape-like, no neck, strong, solid build with long, matted, dark grey-brown fur. It stood behind the tree trunk, which hid its body entirely. I had stopped walking, waiting to see what it would do, as I knew what I was going to do – run as fast as I could in the opposite direction.
>
> "It peered around the tree trunk; I saw no aggression … it seemed intrigued, as nervous of me as I was of it. It then took off to [her] left on a downward slope to the river. It was incredibly fast.
>
> "I stood there amazed, stunned, for about ten seconds, then turned tail and ran, but after about 100 metres I realised it was not chasing me. I was excited [to have] finally seen one, as I [had been] aware of their presence on earlier occasions."

A short while before Lynette's experience, a young friend of the family saw a somewhat smaller yowie quite close to her house.

As readers may remember, the locality of Yowrie (sometimes spelt Yourie) is, in fact, named after the Hairy Man. Up to the 1930s or '40s residents in that region of the NSW south coast hinterland referred to the creatures variously as youries, yowroos or yowies. (Case 167)

Black panthers and UFOs

Lynette also mentioned that she and her husband had encountered a "black panther" on the road, not far from their front gate. Pretty strange – but Lynette then alluded to something even stranger: "UFOs? Don't get me started on them." Without elaborating, she intimated that her property had been visited on several occasions by such things.

So, this story definitely falls into the "high strangeness" category – those cases sceptics automatically consign to the wastepaper basket.

But Tony proceeded to interview a couple of Lynette's neighbours who told of seeing the same, or a similar, black panther on their properties. It seems significant that large black felids have been seen on many other occasions in sites noted for yowie activity in widely separated parts of the country.

Also, improbable as it may seem, an apparent hairy giant/black panther/UFO nexus has also been noted on many occasions in North America.

Case 226. 2000. Malanda, on the Atherton Tableland, about 30 km north of Ravenshoe, QLD

June and Eric Holdcroft

Paul Cropper interviewed the witnesses, June and Eric Holdcroft on 15 October 2018, when they were both 85 years of age. Eighteen years earlier, they'd encountered a yowie while driving, with their granddaughter, from Malanda to Ravenshoe:

> "It was about 2–3 pm, a bright afternoon. It's mainly dairy farms, with forest here and there.
>
> "We were going through the forest when it appeared, less than 100 metres ahead, on the right-hand side of the road, on a slight bend. But as we got closer, we could see this wasn't any normal man, by any means. It would have been about six foot five inches tall, broad shouldered, very hairy, with long arms. It was dark brown and resembled a gorilla.
>
> "It definitely didn't want to be seen. When he saw the car, he looked at us, hesitated, then took off across the road into the bush. By that time, we were quite a bit nearer.
>
> "As he went across, he lent forward; did [sort of] a hop, skip and a jump. It wasn't a run like a normal person's – his left leg seemed to 'hop', somehow. It was most peculiar. He shot across, and all of us said at once, 'What's that?' It only lasted ten-to-twenty seconds, at most.
>
> "When we got to the spot where he went in, it seemed unfathomable that he could have gone through such thick, dense bush – there was no opening – it looked almost impossible for him to disappear just like that, but he did. We kept on driving, a little flabbergasted.
>
> "Our granddaughter was about nine or ten at the time. We spoke to her this morning. She saw it perfectly, and gives the exact same details as we're giving you. She said it was doing kangaroo-like hops across the road.
>
> "We had never heard of a yowie – hadn't any idea. When we got to Ravenshoe and said what we had seen, my daughter's partner said, 'Oh – that's a yowie!' He's lived up here on the Tableland all his life.
>
> "It was all very interesting", June concluded, with masterly understatement, "I remember it like it was yesterday."

Many other sightings have occurred on the Atherton Tableland and in the jungle between the Tableland and the coast.

In addition, an apparent yowie bed was discovered in the depths of that jungle, about 20 kilometres east of Malanda, by one of Australia's most well-respected bushmen, Major Les Hiddens, and a party of scientists, in the 1980s.

Rectangular, over a metre long and a metre wide, it was constructed of calamus fronds that had been bitten through – not cut. It had, said Major Hiddens, "… been slept on that very night … [it] left us quite baffled. Dr. John Campbell, our expedition archaeologist, said, 'If I were anywhere else, I would have to say that was a primate nest.'" (*The Yowie*, pp. 153-54)

Case 227. June or July 2000. Near Gogango, QLD

The witness, Ann Wilton, contacted AYR on 13 July 2016 and was interviewed by Paul Cropper two days later.

The incident occurred as Ann was driving west along the Capricorn Highway with her eight-year-old son, her newborn baby and 12-year-old niece. They'd left Gladstone very early that winter's morning because she had to get to the village of Blackwater by 8 am.

Just after they passed the township of Westwood and were approaching the Gogango Range, her niece pointed towards the driver's side of the road and exclaimed, "What is *that*?"

A large, bipedal creature was running parallel with them, through open paddocks, just loping along: "faster than a stride – a bit like a jog. We watched it for about 10 seconds. It was very early morning but there was good light."

The creature was 7 to 8 feet tall, quite broad, "big, solid," and seemed to have no neck. It had "long, orange hair all over its body – arms legs, all over."

Ann is adamant that it couldn't have been a person in a gorilla suit:

> "It was big enough and close enough [to see detail] but not so close that it could have grabbed us.
>
> "My niece and son were scared then. I could barely peel my eyes away from this thing, to watch where I was driving." Because the kids were "freaking out," she didn't stop.
>
> "A friend said to me later, 'Don't you know? That's the Hairy Man!' But I'd never heard of it. I wish we'd had a camera, but mobile phones with cameras, weren't around then. I'll never forget it – never, *ever*. I'd love to see it again. It was *wild*!"

Case 228. 2003. Jiggi Valley, about 15 km north-west of Lismore, NSW

One of the witnesses, Margaret McRae, reported the sighting to AYR on 17 Jan 2017, and Paul Cropper interviewed her the following day.

"We moved to the Lismore area in 2000 and bought a property in Jiggi. My daughter and I used to take the horses up the hills and all sorts of places in the bush – because I'm a bushie myself. We'd ride for miles."

On the day of the sighting, they rode up Abbey Road, a very steep climb, until they reached Gwynne Road, then passed through a farmer's gate.

"We were walking the horses when they just stopped and put their heads up. And we looked to see what they were looking at and Sally said, 'Mum – it's a Hairy Man!' And it was! And he was really close – just about 20 metres away – and the horses didn't panic.

"[This was] open dairy country – all pasture except for a very small grove of stringy barks and the occasional shade tree, so I was surprised we saw him [there]. But there's dense bush not far away at Nimbin Rocks, and all around there. It's mountainous, steep. Some people who live there don't even walk around their properties because it's so steep. Lots of gullies – a good place for someone like him to hide. [But] he was out in the pasture all on his own."

Margaret McRae's sketch

After a moment the creature broke into a run, but, strangely, didn't run directly away from them. Instead, it zig-zagged: "about 40 metres, turned, then 40 metres back, turn, 40 metres, turn again. He was swinging his arms and legs in perfect rhythm, like a balanced athlete running in a 400-metre race – about that speed. I didn't feel like he was afraid of us – I felt he was showing himself to us, so did Sally. It was really weird. Sounds crazy, doesn't it?

"He zigzagged in front of us three to five times … drawing away … and we got into a trot, and he was never more than 40 metres ahead of us, and then went into that very small grove of stringy barks. And I said to Sally, 'Let's ride up to see if we can see him.'

"This clump of stringy barks was only about five metres square, but [when they got there] he wasn't there. Nothing – just *gone*. We couldn't believe it!"

Paul: "When you say 'gone'… gone where?"

Margaret: "We couldn't see him – I don't know. Vanished! It was a cleared area on top of a hill, and he had nowhere to go except to those stringy barks, and we rode pretty boldly up there – I wasn't afraid – and we got there, and he'd *vanished*. I can't explain it – he wasn't leaning up against a tree to hide himself. The trees – saplings – were only about 20 centimetres across. There was no lantana – nothing – he had nowhere to hide. The grove was tiny, it really was – the size of someone's lounge room.

"But I felt like he *wanted* us to see him – because he could have just run off, downhill and hid in some lantana. But [instead] he ran in front of us.

"It was something I'll never forget. I was looking at the [AYR] website and there's a lot there about yowies but nothing about the [little] Hairy Man. This little chap was nothing like the yowies – he was erect. I said to Sally, 'He runs like Cathy Freeman!' He was athletic – rhythmic strides – the way he was moving his arms – perfectly balanced. He moved *beautifully*. And he wasn't running as fast as he could – he had a speed up, but it was rhythmic and *beautiful* to watch.

"He was only about four foot ten, five foot at the most … 40 to 45 kilos – very, very *slight*. I'd compare him to a Grade 5 student with very fine bone structure, you know – a 'petite' build.

"I couldn't see ears or eyes, no nose sticking out; he must have had quite a flat face. I couldn't see his mouth, so he must've been running effortlessly, breathing through his nose – I'm just guessing. He definitely had a neck. Not long necked, or no neck – just normal. His shoulders weren't wide – a completely lean body. His arms seemed long, maybe because he was so slim. I can't say what his hands were like. I imagine he must have had them clenched – I didn't see any fingers.

"His hair was red – that deep, auburn colour, like a red setter dog. Liver

Chestnut horse colour, and he could have blended well with the stringy barks. I'd say the hair was about two centimetres long and his whole body was hairy."

Paul: "What sort of 'vibe' did you got – neutral? scared? shocked?"

Margaret: "I just thought, 'They're real!' I'd heard about yowies, but I'd never heard about [little] Hairy Men; it wasn't one of my interests, so it never entered my mind. And when we saw one, I'm like: 'Wow – we're so lucky to see this!' It was so awesome. I wasn't afraid, the horses weren't afraid – not at all – and horses can be easily spooked."

Paul: "Did you hear or smell anything?"

Margaret: "No, nothing – he didn't make a single sound. And there was no scent.

"We were new in the area. I mentioned [the incident] to a couple of people and they thought, 'Ah – you've been on the weed!' I don't smoke.

"Sally had a friend, Merlin, [exotic names are common in the Nimbin area] who was living about a kilometre away [from the site of the incident] and he's a really staunch bloke, a mechanic who lived quietly and kept to himself. And he told me that he'd seen the 'Hairy Man' and although I didn't give him a description, he described just what we'd seen. His dog had been barking at night and food had been going missing from the refrigerator out in his workshop, and he went out with a spotlight, and he saw this little Hairy Man. He said an old bloke on a small acreage next to his had also seen it hanging around.

"I spoke to Gilbert Laurie. He's the Aboriginal Elder of the Nimbin area – Bundjalung Aboriginals – and he said, 'Aaah – Hairy Man – I wish I could see one of those little buggers!' He has lived here all his life and he wants to see one too. Gilbert gets quite a few reports, but some people keep it to themselves because you get called a nut."

It's interesting that Margaret's sighting occurred so close to Nimbin Rocks. According to a 1977 addition of *The Richmond River Historical Society Bulletin*, the town of Nimbin was named after small, hairy creatures that Aborigines said lived in the area. They were described as "sort of hobbits." The local Aboriginal term is sometimes spelt *njmbin or nimbinji.*

As mentioned in the Introduction, Aboriginal people in many other parts of the country believe there are two types of Hairy Men: the giant yowie (aka *doolagarl*, *nooncoonah*, etc.) and a much smaller variety known by many names including *junjudee, dinderi, waaki* and *net-net.* Some Indigenous people say the "little fellas" have magical powers, rather like the fairies of Europe.

Margaret isn't the only eyewitness to tell of a yowie inexplicably "vanishing". (e.g., Case 216)

Case 229. 2004. Lithgow, NSW

The witness, Glenys Devine, contacted AYR on 18 December 2014 and was interviewed by Paul Cropper the on same day.

An excellent, lucid witness with a fine sense of humour, Glenys was motivated to report her sighting because she'd "had a minor stroke a couple of days ago, and decided, 'If I'm going to say anything, I'm going to say it now.'

> "I know what I saw, not that I'd expect others to believe without seeing. I'm concerned for [the yowies'] protection and knowing what hunters are like, I don't want anyone to hurt this creature – it had no intention of hurting me."

The incident occurred near the intersection of Bells Road, on the outskirts of Lithgow, and a dirt road leading up towards the Zigzag Railway.

> "It's great bushwalking, all canyons and so forth. I used to go up that dirt road to a particular spot, a natural amphitheatre, and sing. The dog would sit there next to me and howl! I think it was probably the singing that drew [the creature] out.
>
> "So, this day, as we were coming back down the track … probably only 100 metres from the houses, basically at the intersection with Bells Road, the dog – she always walked ahead of me – stopped and turned and was tilting her head from side to side. There had been reports of big cats [the so-called "Lithgow Panther"] in the paper at that time and I thought, 'Oh no, don't tell me it's a big cat!' And just slowly turned around.
>
> "This thing was probably 30 to 40 metres away. There's a great big boulder there which is now covered in foliage, but then the trees were only very small. And I just couldn't believe it – this great, big, huge thing – perfectly balanced, just strolling around the boulder on the rock ledge. I watched it from side-on and rear view for probably 20-odd seconds, as it went behind the rock.
>
> "Once it went behind the rock I didn't hear or see another thing – so I presumed it squatted down. [Although it hadn't looked directly at her] I think it was well aware of me, in fact, I think it was following us. Had the dog not stopped, I wouldn't have seen a thing. So, it made me wonder how many other times this thing had seen *me* – I'd been walking there for two years.
>
> "If he wasn't covered in hair [he would have been] an absolutely gorgeously shaped guy. He was tall, almost fine-boned, but had what us girls would call a 'desirable rear shape!' [Laughs]
>
> "His legs were long, and his arms were maybe just a little longer [than a human's] but very well proportioned. I could see its hands. I know that sounds

silly because it's only an animal, but it really was like a human, but taller, and covered in beautiful black hair, thick on the shoulders. It was about ten centimetres long, denser on some parts than others but I could see through it and see the muscles working.

"You're thinking, when you're looking at something like that, 'Could it be someone in a suit?' [But] I could see the muscles working in the calves – and that was no suit. The skin was almost black, and the hair was jet black. The feet [were hairy] as well, but I could see its toes.

"I saw him from the side and the rear, but I couldn't see the face – there was hair all down the side of the face – I couldn't see his nose sticking out, for example. The hair on the head was thicker than on the legs; I think the ears were probably covered with hair.

"It had a proper neck. It wasn't hunched over, it was like a guy that's really fit before they take on steroids, if you know what I mean – muscled but not over-muscled." The shoulders tapered to the waist in a vee shape. "Fairly narrow hips, but the legs were muscled. If he was shaved, you probably wouldn't think he wasn't human.

"He was absolutely beautiful. And although I didn't see it face-on, because of its physique, I thought it was a male. Nothing feminine about it. So, I figured, 'Well, that's a young animal – a teenaged animal.' But I wasn't scared. I didn't smell anything – the wind was probably going the other way.

"It was *bold* – I really feel it wasn't scared of me in any way and I wasn't *truly* scared of it – but it scared me enough [laughs] to not go back there for two weeks!

"We stood there a minute or two longer and I had my hand on my heart. And then we kept walking. We kept glancing back but I didn't hear or see another thing. On the way home, I bumped into a neighbour, and I didn't know whether to say anything or not. Anyway, I told her, and she wasn't surprised. She hadn't seen anything herself, but she'd heard many, many stories."

When Glenys returned to the site two weeks later, she was amazed to find that the rock ledge the yowie had negotiated so easily was two metres from the ground and only 5 to 10 centimetres wide.

"When I climbed on the boulder, there was no way I could have strolled around [the ledge]. I had my feet splayed out and was just sort of inching my way around, clinging, face against the rock – and I used to do a lot of rock climbing back then. But this thing *strolled* like it was walking on the road – its balance was unbelievable.

"Looking back, I can't see any reason why he was walking that ledge – except to show off!" [Laughs]

Using the height of the boulder as a guide, she estimated the creature was about seven to eight feet tall.

"At the time, I couldn't believe what I was seeing, so I was trying to think logically. I never once thought it was human – I knew it was an animal. Even the dog didn't think it was human – tilting her head from side to side.

"I consider myself very lucky and very blessed to have seen it."

It's interesting how similar, in some ways, Glenys' report is to that of Margaret McRae. (Case 288, above)

Although they were very different in size and colour, both women thought the "perfectly balanced", fine-boned, "beautiful" male creatures they encountered looked like teenagers. Glenys' creature was "bold … show[ing] off." Margaret's was "showing himself to us."

Case 230. c. 2004. About 15 km east of Jimbour, QLD

One of the witnesses, Irene McLucas, contacted AYR on 26 March 2014 and was interviewed by Paul Cropper on 28 April.

"It was summertime, I was chipping some cotton and staying with my sister in a little farmhouse on the Jimbour to Bell road.

"My sister, myself and my friend were in the front of this little truck, heading home about 8 o'clock [through] open, scrubby country. We had our headlights on, came around a corner, and caught this unusual being standing up – upright.

"It looked like he was going to run across the road, but he got caught in the lights …. He had his arm up on a tree, like he was going to use it as leverage to propel himself across the road [but] he just stood there, like he was stunned from the lights.

"The hair was kind of long and shaggy – a bit like a German Shepherd's hair, and the eyes came up a yellow colour from the car's lights. It wasn't very tall – about four or five foot. My mum is five foot, and it was around the same height as her.

"We weren't driving that fast. It was just standing there looking at us. We were all just stunned, no one said a single word – we just stared at it, and when we were past, I said to my sister, 'Let's go back!' And she said, 'No way!' And put her foot on the accelerator and took off like a maniac. Then I said, 'Did you see what I saw?' And she said, 'I don't want to talk about it!' And the other lady didn't say anything.

"So, when we got home [and the sister still refused to talk about it] I said, 'Alright, just draw a picture of what you saw, please.' And we all went into separate rooms, and all drew basically the same picture."

Paul: "How long was the creature in view?"

Irene: "Oh, probably only 30 seconds or so."

Paul: "How close were you at the closest point?"

Irene: "Probably about 15 feet away, pretty close – it was right on the side of the road. I was sitting on that side, so I got the best view."

Paul: "Was it built like a person – a man?"

Irene: "Yeah, I guess so – it was upright and had arms and legs. The arm looked longer [than a human arm]. It was just a normal build. It was completely in proportion [and looked] very agile.

"Long, shaggy hair, say about four inches long all over its body. A tanny or earthy colour. The ground out there is a reddy, sandy colour, and it was kind of like that. It was at night, but that's what the colour looked like.

"The eyes [reflected] the light, so you could see where they were, but you couldn't really see the nose and mouth properly – I think the hair was kind of covering it."

Paul: "Could it have been a person in a suit?"

Irene: "No – no way in the world – not out there – definitely not. There's no doubt about it. I was getting the sense that I was looking at a critter I've never seen before. There was no mistake: it couldn't have been anything else."

Case 231. 2006. Colo Heights, NSW

Alison Wilkins contacted AYR on 23 August 2013 and was interviewed by Paul Cropper on 13 January 2014.

The sighting occurred, shortly after she, her partner, her sister and her two sons set out from Colo Heights at around 11 o'clock one night, to drive towards Singleton:

> "My partner was driving, and I was beside him. We were having a conversation as we came around a bend … just outside Colo Heights … heading towards the service station [that is situated just before the turn off to Singleton] and we seen something on the road – on the right-hand side.
>
> "It was a fair way ahead, but we seen it pretty well because of the headlights. It looked at us and you could see that it was walking on two legs and was completely covered in hair.

"It walked across the road pretty fast, with big steps, and went behind a big tree, but kept looking out ... like it was holding onto the tree and peeping out. And as we were about to go past, it looked out again and we got a pretty good look at it.

"It didn't look much taller than me [about five foot six] and was a little bit slumped over and covered all over in grey, silvery hair, about 15 centimetres in length. The hair was fairly straight – not curly or wavy.

"You could see, like, muscle structure ... arms and hands and head. You could see it had big eyes and a nose and mouth, but I couldn't say what colour its skin was [because] it had a fair bit of hair on its facial area.

"It wasn't [the face of any animal], it was like a cross between a human's and a gorilla's face. Its eyes were bigger than a human's; when the headlights shone on them, they were a yellowy colour – like a cat's eyes in the dark, I suppose.

"We kept driving and were all silent – except my hair was all standing up on the back of my neck and I was really scared. And I said to my partner, 'Did you see what I just saw?' And he said, 'Yeah – I saw it too – what the hell was it?' And I said, 'Was that a yowie?' And he said, 'That's what it looked like – I've got goose bumps and my hair is standing up!'

"And my sister was freaking off her head because she'd seen it too and she's going, 'But yowies aren't real! It's all made up! It can't be a yowie!' And I said, 'Well – what else could it be? Nothing else walks on two legs and it wasn't a human.' Anyway, any ordinary animal would have just kept going after crossing the road – not stopped and looked to see if we were still coming.

"So, our first reaction was, 'Oh f*** – it's a yowie!' And I can't even describe to you this fear that you get when you see these things. I've been scared before, but this was way beyond that. It's like nothing I've felt before. It's a different kind of fear you get – that goes through your body.

"Like my partner said, 'We were in a car doing 100 kmh, so what are we scared for? It can't get us!'

"It puts the fear into you – that's the scary thing. I even said to my partner I don't want to ever go back there!' We just wanted to get out of there – we were petrified.

"Anyway, we got to the service station and pulled up, and my partner talked to the bloke there. [And the man had a lot to say about yowies frequenting the locality.]

"I don't wish to see another one. If it's dead, I'll [look at it] – but not a live one! [Laughs] I've always been one of those people: 'I'll believe it when I see it.' Well, now that I've seen it, there's no doubt in my mind they exist!"

Determined sceptics might suggest that Alison and her family were fooled by a clever hoaxer in a gorilla suit equipped with reflectors for "eyes." But because her vivid description of the overwhelming, unreasoning fear that they all experienced – even while safely locked in a moving car – tallies so well with the "nameless dread" described by so many other witnesses throughout the decades, we believe this was a genuine yowie encounter.

Case 232. 6 Feb 2006. Springbrook National Park, QLD

This report came to us courtesy of naturalist Gary Opit:

> "Matthew Schwerdtfeger and Shaun Smith informed me that a large rock was thrown at them from the forest near the Springbrook Repeater Station, behind the Gold Coast, at 4.30 pm on 17 September 2005.
>
> "Then, on 6 February 2006, a black-haired, man-like animal threw seven large rocks at them from the forest while they were standing on the acre of cleared grassland between the communication tower and the forest edge."

Immediately after the volley of rocks, the duo took a photograph of the scrub from which it emanated. They believe that the resulting picture reveals partial silhouettes of two yowies amongst the foliage, and maybe they're right – they were, after all, right there, on the spot, being pelted with stones – but to us the shot is too shadowy to prove anything.

That day, they also took photos of large, flat, five-toed footprints. Although those, too, were indistinct, the story is still very interesting, because the spot where the action occurred was only 100 metres north of "Best of all" lookout, where ranger Percy Window experienced an extremely close yowie encounter in March 1978. (*The Yowie*, pp. 2-3)

Another thing that suggests Matthew and Shaun really encountered yowies that day is that another rock-throwing incident happened in virtually the same spot eight months later. (Case 233, below)

It is also worth noting that all four of those incidents occurred close to the massively powerful Springbrook Repeater Station. We have noticed that electrical installations and powerlines – and lightning storms for that matter – are frequently mentioned in yowie reports.

Case 233. 2 Sept 2006. Springbrook National Park, QLD

This story also came to us courtesy of Gary Opit.

On 2 September 2006, Andre Clayden, manager of the Springbrook Homestead Restaurant, phoned Gary to say he'd just received an interesting report from a Chinese tour guide who regularly brought tourists to dine at the café.

The man asked Andre if there could be someone in the area who didn't like tourists, because, as he was leading his party back from "Best of all" lookout along the rainforest track, someone began screaming at them and throwing rocks. He said the voice sounded too powerful to be human, and the figure they saw through the trees looked too big and hairy to be a man, but he didn't know what else it could be, as he had not heard of the yowie.

As mentioned above (Case 232), the area around "Best of all" Lookout – seems to be something of a yowie hot spot.

Case 234. 19 Sept 2006. Megalong Valley, Blue Mountains, NSW

Catherine Bolton and Sarah Baikie contacted AYR the morning after the incident and were interviewed by Paul Cropper three days later.

While horse riding through rugged country in the Megalong Valley, the friends experienced repeated close encounters with a small, hair-covered hominid.

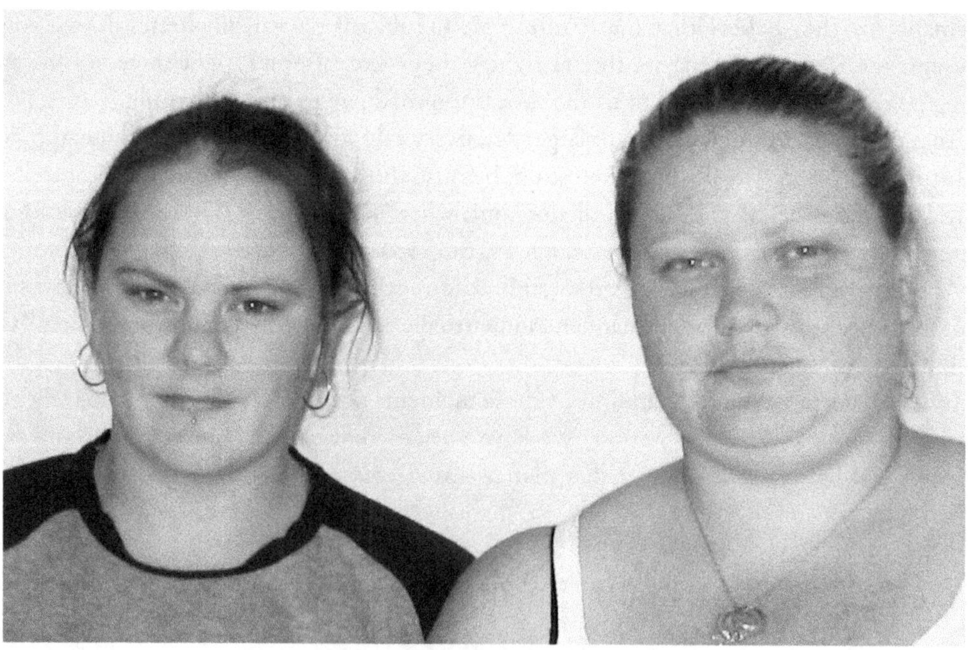

Sarah Baikie and Catherine Bolton

Catherine: "I'm a pretty experienced rider. There was me, my husband Brendan, Sarah, another girl we didn't know, and the trail guide [from Werriberri Trail Rides]. We started about 1:15 pm.

"We were about 5 to 10 minutes into the ride; the others were 20 to 30 metres ahead when my horse stopped and refused to move. He was sniffing the air and turning around trying to bite me [so] I knew something was wrong. Then I smelt, like, salty blood – a real foul stench – and I looked around to my left and it was right there!

"It looked sort of like a monkey, but more human – a lot smaller than a person – about four foot. It was solid – square shoulders – pretty much like a square block. Very hairy, dark brown, all tangled, like a shaggy dog that hadn't been washed for a while, and mud all over it ... long hair everywhere. I had no idea what I was looking at – I pretty much crapped my pants!

"It had a pushed-in nose. I distinctly remember two canine teeth out the front, outside the lip. I couldn't see ears because the long hair covered them. I saw eyes, but not distinctly.

"I saw arms, and I only saw three fingers, but there could have been more, because it had something in its hand – something like a dead kangaroo, but smaller. There was [raw] flesh, like whatever it was holding was skinned – like it was inside out.

"The bush was quite open [at that point]. He was 10, 15 metres away, if that. I was looking down at it and could see right down to its feet – I could see everything. The legs were long, but he only had three claws on his feet. It just stood there and looked at me ... It seemed like forever ... but it was only two or three minutes, if that.

"I kept kicking the horse and held on tight, and it took the track pretty quick."

As her horse began to overtake the others, Catherine could still smell the strange creature, and, although she could no longer see it, "I felt like it was watching me. I held on for dear life and caught up with the others, then all of a sudden we heard the sound of a branch breaking behind us."

As her horse attempted to pass the others, Catherine fell off, still holding the reins, and pulled the horse down. The trail guide asked her if she'd like to go back to the stables, "But I said no – I didn't want to wreck their day. Then, about another half hour further along – I was freaked the whole time – I kept smelling the foul smell. Brendan also smelt it then."

As they proceeded, friend Sarah was just ahead of Catherine, and Brendan ahead of Sarah. "Then it [the creature] moved past us and Sarah said, 'What the f*** was that?' And I caught a glimpse of the back of it [as it ran beside the trial] and my horse took off, flew past the other two horses and I hit a tree with my shoulder, came off and hit the ground really bad."

This time Catherine was seriously injured – she had abrasions on her right forearm and hip, fractured right collarbone, two fractured ribs, bruised legs, and swollen ankles. "I was in pain, bad. My horse ran off and I was taken back to the stables on Brendan's horse, and he drove me to Katoomba Hospital."

Sarah fully corroborated Catherine's story.

> **Sarah:** "Before I saw it, I smelt something like dried blood ... sweat, sort of ... then something moved ... I heard a little scuffling in the bush and then I saw it. It looked like a monkey ... an ape sort of thing ... hairy.
>
> "There were trees and rocks and little bushes here and there [so she saw only the head, shoulders, and upper chest]. It just popped out, stared at me, and then ran. And I went, 'Whoa!' And Cate said, 'Did you see that?' And I said, 'Yeah – what the f*** is it?'
>
> "It wasn't big [about four feet tall]. A browny, blacky colour ... all long, scruffy hair. Half human, I reckon, by the looks of it – all hairy, but human-looking. I focused on the eyes and the mouth. Mouth sort of half open. You couldn't miss the teeth – they were about an inch long, over the lip.
>
> "It's like it was trying to target us. I was kicking my horse – I just wanted to get out of there! It's like, 'Is this thing is gonna eat me?' My adrenaline went ... I could have run faster than the horse!"

On her release from hospital Catherine phoned Werriberri Camp to ask after her horse: "It had come back all shaken up and had to be rested. The owner said this had only happened once before – about seven or eight years ago [c. 1998/99] when a group of very experienced riders had come back, all as white as ghosts. They'd seen and smelt the same thing."

Case 235. January 2007. Howard, 30 km north of Maryborough, QLD

Wendy and Zai Ericson contacted AYR in January 2007 and were interviewed, on site, by Paul Cropper and a retired policeman (who requests anonymity) a few days later.

At the time of their yowie experiences, Wendy and 13-year-old Zai were living on the outskirts of the village, across the road from two 40-acre properties. Beyond those lay a large expanse of state forest.

Zai and Wendy Ericson

Wendy: "We used to go riding on the neighbouring properties – Mick Howard's and Steley's – but we never liked Steley's, and eventually avoided it [because it had] a 'haunted' vibe – a really strange energy, the horses don't like being in there. Sometimes you go in there and there's no birds, no insects – deathly quiet – really creepy.

"Roughly two weeks ago we decided we'd exercise the horses, take them through Steley's, then through Mick's and back – a nice long ride.

"The first thing that caught my eye was two trees broken in the same direction, at the same height, and I said, 'Look at this – it looks like an elephant has walked over the top of them!'

"And then we noticed trees down everywhere – pulled out by the roots and branches broken, and I thought, 'This is a bit weird', and the horses were getting a bit 'tichy.' My daughter's horse just wanted out of there!

"So, we came back, onto Mick's property, and were talking about what we'd seen, and then Zai looked down and saw this print and said, 'Look mum – there's a bear in here!' And I looked down and thought, 'That's not right – that's something weird'."

Zai: I thought, 'a bear!' Because I couldn't think of anything else. It had the normal toes like a human's foot, and the impression of the heel and there was a fourth toe there – down on the bottom at the side, and I said to mum, 'Is that a bear?' And she says, 'Hmmm – *no*'."

Wendy: "It was, like, humanoid. And I thought, 'I think I know what this is!' So [when they got home] I contacted the Australian Yowie Research site."

At the suggestion of Dean Harrison of AYR, Wendy bought some plaster of

Paris and at about 4.30 pm the following day she, Zai, and her niece Stephie, rode to the spot to cast the track.

Wendy: "It was such a *big* print that I eventually used the whole bag of plaster, but after I started mixing it, I realised I hadn't brought enough water, so I asked Zai to go down to the dam for more. She was gone for a while and then comes tearing back with her arm up in the air, and I'm thinking, 'Oh God, what's happened?'"

Zai: "I galloped to the dam … got off my horse, grabbed the water bottle out of the saddle bags, bent down, and filled it up [and then] heard this strange little noise over to my right. It was like a, 'Hih, he, he, he, he!' I'd heard it before, but I can't remember where.

"So, I put the bottle in the saddle bags, and my horse spun around as I got on, and I looked over and saw a little, jet-black, fuzzy creature – it was only about four foot high – just standing there with its hands like this: [Hands together, palms outwards, concealing most of the left-hand side of its face]".

The little creature was very close – only five-to-ten metres away.

"All I could see of the face were its eyes, the size of twenty-cent pieces, and they were just black – I couldn't see any whites in them. The face was just fairly fluffy. [She could see most of its body] but I couldn't really tell which part was which [because of the fluffy, eight-inch-long hair that completely covered it.]

"Very fuzzy – strange. But I just thought, "OK, if this is the baby, where's the mum? [Laughs] It didn't move at all. I only saw it for a couple seconds – I wasn't going to hang around! I turned my horse and took off at a gallop, up the trail. And then I looked – and the saddle bags had fallen off at some point!"

Wendy: "When she reached us, I thought she was going to throw up – she was *grey*, and she said, 'I've lost the saddle bags!' And before I could stop her, she'd spun the horse around and she's off – back to get them!"

Zai: "I saw the saddle bags on the dirt road at the bottom of the hill, and a couple of metres before we got to them my horse just *stopped* – he wasn't going any further. And he just turned his head and looked over to the right, and I looked and saw this really dark, black figure leaning over like that: [Body side on, arms extended at shoulder level, hands clasped together, hands against a tree, head turned to the left – towards her.]

"And I thought, 'Oh, it's just part of the tree – a burnt limb. But – *no!* It had shortish [about two inches long], shiny fur, not fluffy like the baby's.

"So I thought, 'Well, we need that water bottle – we need the evidence [the footprint cast] and I just slowly dismounted, and it didn't move – it just watched me. I grabbed the saddle bags and when I got back on, I looked again, and he must have realised I'd seen him, because it put its hands down to its side and turned around to face me ... full-on to the front.

"It was taller than my uncle, who is six foot three, so I'd put it about seven, eight foot – it was very big. When [the retired policeman] walked over to the same tree a few days later it was *way* bigger than him. [Indicates two to three feet taller.]

"I could see it was very well-shaped around the arms and you could see the muscle on its stomach. And its shoulders were like [indicates three to four feet wide].

Wendy's sketch

"[What] really got to me were the two little white teeth [protruding] over the bottom lip. There was absolutely no hair from the top of the chest to the bottom of its stomach. The skin looked grey, just grey.

"It was just silent – I didn't even hear the leaves crunching when it changed position … a bit like an apparition, in a way … a bit like a special effect on a tv show.

"I didn't smell anything. It just looked at me and its eyes were bigger than the [smaller creature's] eyes – about the size of a fifty-cent piece – and they were jet black. [The face] was shaped like a person's because it was round. But it was very flat at the front and one feature I noticed was that its nose was sort of curved, so it almost met back at the [upper] lip. Its features looked human, but the hair and its build was like an ape. It looked like … nothing. I've never seen anything like it before.

"For some reason I just sat there for a couple of minutes, looking at it. And it was, like, studying me. And I thought, 'If I sit here, it'll eat me. Right – it's time to go!'

"I took off, and I thought it would easily catch me up, it was so big. And I heard, beside me, a couple of strides like, crash, crash, crash – and I didn't hear anything after that.

"When I got back [to her mum] I felt really sick – nauseous – and I said, 'I saw another one – it was really big!' And she said, 'We'll just do this cast before we go.' And her horse was carrying on."

Wendy finished pouring the plaster. (Unfortunately, her efforts were wasted – when the cast was retrieved later it was too malformed to be of any use.)

> "We headed off and mum says, 'Oh, we've gotta go back and get a picture of it.' And I said, 'You're not going anywhere – not with something that big hanging around!'"
>
> **Wendy:** "We were riding at a quick walk, trying not to look as if we were frightened, and Zai kept saying, 'Don't trot, don't trot!' My horse was playing up something shocking. My niece's horse was also zigzagging – he didn't want to be there; Zai's horse is older and braver.
>
> "Zai was telling us what she'd seen earlier, and then she turned in the saddle and said, 'As a matter of fact, it's watching us now … over there, near that tree.'"

The tree was quite close – about 30 metres away.

> "I turned around to my left, saw it, and said, 'Oh f***ing hell – let's get out of here!' It was facing a tree and had its arms up like that [see sketch]. But [then] it turned its head to watch us and the distinct impression I got was that it was telling us, 'You keep going through and I'll just stand here.'
>
> "So we rode on, and it just stayed still [watching them all the way]. We had it in view for easily five minutes. To my eyes it was just jet-black; I could see the shine off the hair – very sleek. And [although the sun was behind it] it was solid, three-dimensional. [It was partially side on] but I could see its shoulder shape. What struck me was the width of the shoulders. [Indicates a shoulder width of about four feet.]
>
> "It was like a humongous gorilla shape. I saw a head shape – I saw its head move – it looked a little bit small for the body. I could see from the head down to about the waist. It definitely had arms. It was probably eight or nine feet tall. Muscly – the whole impression was strength.
>
> "By the time we got to the gate it was getting to be dusk and the animal was still there. We came straight home and by then it was dark, and I heard some really loud yells from over at Mick's property. These were *really loud* yells – like a pig's squeal and a roar and an 'Oop, oop!' And I'm saying to the girls, 'Listen to that!'"

"High strangeness"

At this point the episode, already very strange, began to take on elements of "high strangeness" – the girls couldn't seem to hear the vocalisations. Another thing Wendy found puzzling was her niece Stephie's reaction – or lack of reaction – during their yowie sighting.

"I was so surprised. She was so calm – looking at this thing as if she was just looking at a tree frog or something – but she had a big reaction later. That night she got a migraine headache and really bad diarrhoea and the next morning she was sick and said she's never going back in there again.

"I had a headache too, that night, and a worse one when I woke up in the morning. Zai had nausea and a headache, too.

"The next week, we took [the retired policeman and another man] and showed them the damaged trees and where the print was. They heard some really loud, rhythmic banging. I heard it later on – it was like a branch being hit against a tree trunk – loud, a definite bang; I heard a couple [of repetitions]."

Wendy adamantly rejected the notion that she and the girls could have been hoaxed by a person in a gorilla suit:

"Oh God, no! Too big. And considering what Zai saw, too elaborate anyway. And Mick Howard had padlocked all his gates – we were the only ones allowed in."

A week or two before the yowie encounters, Zai told her mother that she'd experienced a couple of less dramatic – but nevertheless rather odd – occurrences while riding in the area:

"She said things were being dropped on her helmet, and I thought it must have been stuff out of the trees, or birds. But she said there were stones and berries and sticks – this was over a period of time."

Stranger still

Like a significant number of other yowie witnesses, both Wendy and Zai mentioned several other very odd, apparently psychic, events they'd experienced throughout their lives:

Wendy: "This [the yowie incident] was *one* of the most unusual experiences for me, and *the* most unusual for Zai. I've had unusual experiences all my life, since I was a little girl.

"The earliest I can remember was seeing fairies in the bedroom when I was about six, and that was a definite sighting, quite real. And the [actions] they were doing … It wasn't until about 25 years later, when I researched [faery lore] that I realised it was a travelling troop – because they had little carts ….

"As I grew up, I've studied religion and spirituality."

Zai: "I've only seen one other thing that was equal to [the yowie event]. I was in Burrum Heads, about 10 minutes from Howard, walking my Staffie along the beach – I think it was midday – she loved the beach, and we were about 3 to 4 Ks from where you enter the beach.

"All of a sudden, she almost ripped my arm off – she wanted to go back home. She'd never, ever, done that before, and I looked up [to where] the mangroves started to grow along the shore, and there was this really tall, red figure. Bright red – like there was sort of an aura about it.

"It would have had to be 10 foot tall, easily … really lanky and tall – straight up. It was basically all legs, and its arms were too long for its body [they] finished around its knees, and its torso was probably only about that big [indicates one metre]. And it was just standing there, looking directly at me.

"A humanoid shape – but it was *red* – the actual skin colour was red. I couldn't see any facial features; it looked like [laughs] play-dough – basically no face or anything.

"It just stood there, and I just really freaked out. It was 100 metres away, there was no scrub or anything, so I could clearly see, and I was like, 'Okay, okay – I'm going!' And my Staffie dragged me all the way home."

As she and her dog reached the exit from the beach, "I looked over to the big curve of the beach with the mangroves, and you could see this bright red aura – a glow like when you see a fire from a distance away."

With Wendy and others, Zai has also seen "quite a lot of UFOs":

"One time [when they lived in the village of Kenilworth] we came home from Gympie, driving along a dirt road, and a really fluorescent, green light caught my eye, up in the sky. I turned around and could clearly see the outline of a really big, grey, round thing. Light was coming out behind it – not like flames – like globby shapes of lights.

"And it crashed into the side of this large mountain, but there was no sound – not a sound. There was a farmhouse just at the bottom of the mountain, and we could see the people just doing what they'd been doing before it crashed – just silence."

A terrifying apparition

"When I was younger, we used to live in a unit on the Sunshine Coast, and I used to wake up – I was only about six – and I used to see this black old woman. She [was wearing] a cape and had a staff, and I used to run past her into mum's room and I could feel the cold as I ran past her. I'd just stand there, holding onto mum, and this thing would come into mum's room and sometimes come so close to me that I could feel the breath on the back of my neck.

"That happened a *lot* of times, so it was definitely not just a figment of my imagination. It only ever happened in that unit."

Where to begin? This is a particularly interesting case, involving two yowies encountered at close range, in broad daylight, by three witnesses.

Wendy and young Zai impressed us as being very competent, unaffected, honest people, and their story was corroborated to some extent by the fact that Paul Cropper and the retired policeman examined the area of broken trees and flattened undergrowth. The policeman also heard the "loud, rhythmic banging" described by Wendy.

Wendy and Zai were not only physically courageous in their interaction with the yowies, but also morally courageous in their willingness to put on record the many other very strange, evidently psychic, events they'd experienced throughout their lives.

The region around Howard has produced several other sightings of hairy ape-men, both large and small. (Cases 115 and 220, and *The Yowie*, p. 126)

Even Zai's sighting of the skinny, bright red, ten-foot-tall creature is not quite unique – businessman Richard Kingsley saw a similar, but jet-black, "stick-man" in the Northern Territory in 1996. (*The Yowie*, p. 271). Zai and Richard's descriptions of the creatures sound very much like the quinkin figures depicted by Aboriginal artists in northern Australia.

Case 236. 1 Jan 2008. Near Coffs Harbour, NSW

Sharron Williams contacted AYR on 19 June 2017 and was interviewed by Tony Healy on 3 December 2017.

Her encounter, with a creature so bizarre that it seemed more of an "alien" than a yowie, occurred on a stretch of the Pacific Highway that passes through Bongil Bongil State Forest, just short of Coffs Harbour. At 2 am on New Year's Day, she was driving alone from Nambucca Heads to the Gold Coast:

"I took a risk – I shouldn't have left at the time I did … but [I was] under duress [and] I was really 'running the gauntlet' because I had very little fuel.

"The area was quite deeply forested, and they'd been changing the highway there – the road had been split, and quite a lot of clearing had been done, so that gave me good visibility, even at night. I still had quite heavy foliage on the left-hand side [although] it was partly cleared.

"I noticed something dark – I thought it might have been the shadow of the vehicle reflecting through the trees. My dog was whimpering [and] I thought,

> 'What's going on?' She never whimpers, she enjoys travelling in the car. She jumped down and curled up under the dashboard – that seemed bizarre.
>
> "'Then the shadow I'd noticed in my peripheral vision seemed to be coming towards me. This thing was coming closer and the forestry line closing in, so the thing [would soon emerge from cover]. It was no longer just a shadow – I could see colour. It looked orange – all this *flowing* hair. I thought, 'That thing's going to come across, and we're going to see what it is!'"

The creature emerged from the tree line, onto the road, and ran alongside her car at a speed impossible for any normal animal – about 95 kilometres per hour.

> "He was on all fours at first, head down, just keeping up with me – as if it was just playing with me, enjoying itself. But it came closer, closer, and I thought, 'Here we go – he's going to cut across me!' And then it stood up on its hind legs and ran in front, right in front of my headlights. I'd sped up to 120 kph and it *still* managed to get in front of me. I thought I was going to roll the car!
>
> "It came in front and *showed itself* – showing it was so powerful. [Then it accelerated again and crossed to the other side of the road]. It *powered* across – it must have been doing 160 – over the other lane, into the bush – and just *gone*!
>
> "I had to pull over to reconcile myself, and the dog was *screaming*.
>
> "He was lean, skinny, with very long, distended arms; very, very long-legged. It was massive: probably over eight feet tall. It had human characteristics – human hands, human feet, human face – like an ape, very dark brown skin, almost chocolate-coloured in the areas I could see.
>
> "And the hair just *flowed* off him. The colour of a red setter. I've since seen [illustrations of yowies and the length of their hair] but this wasn't like that. The hair was *extreme* – it was *so* long – as if I grew my hair from the top of my head down to the ground. It was all over its body [streaming back from its head, shoulders, and arms as it ran.]"

In addition to the bizarre creature's unearthly turn of speed, there were other things, less easily definable, that gave Sharron the strong impression she'd encountered some sort of alien:

> "Because of the burst … the influx of power … it's almost like he shape-shifts – almost that kind of feeling. When it came in front of me it *showed itself* – just totally showed me that it had so much power – more power than the vehicle. When he got in front, he turned his face for one split second – almost to 'smile' at me.

> "It was magnificent, actually. I was in awe. [But] it didn't want to harm me – it was *playing* with me. It was the most incredible thing I've ever seen – and I was *not* dreaming.
>
> "I believe they're out there and that there must be a colony of them, because this creature was young. Imagine if we were up against a family of big ones! [Laughs]
>
> "I've carried this experience with me for a long time without talking about it, and I think there might be a lot of other people who've had similar experiences who never speak about them."

Although Sharron's report was very strange, to say the least, she came across, in conversation, as an entirely open and uncomplicated person.

It's interesting to compare some of the weird red giant's appearance and behaviour to that of the much smaller creature Margaret McRae and her daughter encountered near Jiggi, NSW, in 2003. (Case 228)

The hair of both creatures was "the colour of a red setter", and both ran across in front of the observers: they "wanted to be seen" – were "showing" themselves.

One thing Sharron said (and is willing to have mentioned in this book) might be significant. As she drove the dark highway alone, low on fuel, "running the gauntlet," at two hours past midnight on New Year's Day, she was feeling very stressed – "under duress" because of a domestic situation.

We know of other cases where witnesses were similarly stressed, sometimes because of an argument with their partner, sometimes for other reasons, shortly before encountering a yowie. One such case – quite a remarkable one – is, in fact, included in this collection, but as the witness didn't give us clearance to do so, we deleted all reference to his marital spat.

One amiable fellow, Justin Garlick, who we interviewed for our previous book, was, like Sharron, quite "out front" about the tiff with his partner that immediately preceded his long drawn out, very close-range interaction with three massive Blue Mountains yowies. (*The Yowie*, pp. 115-117)

Case 237. Sept 2008. The Alpine Way, near Khancoban NSW

Valerie (surname confidential) contacted AYR on 29 Aug 2018 and was interviewed by Tony Healy the following day.

A high-ranking health officer whose area of responsibility encompassed much of south-east NSW, Valerie was 48 years old at the time of the incident.

On the day in question, she'd attended a community meeting at Bombala, after which she set out for South Australia via the Snowy Mountains to visit her daughter. She planned to stay overnight at Corryong, and by about 3:30 pm was on the Alpine Way between Geehi and Khancoban.

> "It was a beautiful day, but I was driving *really* slowly – probably only about 30 kilometres an hour – because there were remnants of snow on the side of the road and evidence of rock falls. [That stretch of road passes through] typical, thickish bush. I didn't pass a single car – mine seemed to be the only one on the road.
>
> "Then, when I was about 30 minutes from Khancoban, three kangaroos suddenly bounded across from the right, but because I was driving so slowly, I didn't have to brake – I just took my foot off the accelerator. They came out *at speed*.
>
> "Then, out of the same spot where they had come from, I saw this *creature*. I have never seen anything like it. It was [on the edge of the bush] but practically on the road.
>
> "I thought, 'Good heavens, what on earth is that?' And I was just *looking*, driving slowly. And it was looking at me – right into my eyes; and that's when I had the impression that I was looking at something that had intelligence. And I thought, 'This is not an animal!'
>
> "It was about five feet tall – it's very difficult to judge the height of something when you're driving. It was very solid, but I didn't get the impression it was huge. He looked for all the world like a rugby league player – that kind of a build, but shortish.
>
> "Very thick set, wide, and his head looked almost like it was bolted to his shoulders [laughs] I didn't get any sense of a really good neck. It lifted one arm to push a branch aside. [The arm didn't look disproportionately long.]
>
> "It had a big-ish head and black, straggly hair down to the shoulders. [The rest of the body] was practically all bare. The skin was a pale, grey-white colour … creamy. But it was really quite odd, because I couldn't discern what gender it was [even though it was hairless and naked.] It was just this *creature*."

As it was standing very close to the edge of the road, the creature must have been within about ten metres of Valerie as she drove slowly past.

> "It's very hard to describe something that you've never seen before. I got the impression its facial features were rather flat, but that's just an impression, because I was only looking at him for 15, maybe 30, seconds. I just got a sense of its face – big face, big head – and … it's odd, but I can't describe the ears, nose, lips, or anything – just the eyes.
>
> "He was looking at me as I went past. [Its eyes weren't huge: no larger than

those of a human] and – I'm not sure now if I'm creating this memory after the event – but I thought it had pale eyes.

"It stared at me ... for quite a few seconds – as if he was as curious about me as I was about him. I returned the stare and got a sense that I was looking at an intelligent, sentient being ... it seemed like he was weighing me up, making a decision – like, 'What will I do?' And then he just turned around and walked back into the bush, totally unconcerned – there wasn't any sense that I had startled him in any way, just that he was curious about this human. But no, I didn't stop – you know, a woman on her own – but as I was approaching him, I did lock my car doors! [Laughs]

"It was incredible, but it's funny – I wasn't frightened, just puzzled, because I have never, before or since, seen anything like that. And I know what I saw is real because I am totally *not* delusional.

"It wasn't [any normal animal]. I've been in Australia most of my adult life and I didn't know what it was. I just kept driving, puzzled. I wasn't frightened – but then again, I work in mental health, and nothing much frightens us! [Laughs].

"I couldn't believe what had happened, and when I got to Corryong, I phoned my son and had difficulty explaining what I had seen. I didn't want to call it an animal, because he, she, or it, had looked into my eyes – we made eye contact – and I got a feeling of a sentient being. My son suggested that it might have been a yowie. I believe that I have seen a yowie.

"I could never really describe it to people, but we were watching something on television last weekend about yowie hunters, and there was somebody who had drawn a picture, and I just thought, 'That's what I saw!'

"I am a very rational person, but this experience has rattled me ... it was a shock to realise I'd seen something that isn't supposed to exist. But I know what I've seen!

"Now I've done some reading about yowies and I can't understand how no evidence has been found. I know it's a big country, but I find it incredible that we can't find any real evidence.

"You would think that eventually we would find bones or something like that. I don't believe in the supernatural, I believe in evidence – my line of work is evidence-based. But this experience made me think that maybe there are more things in heaven and earth than we have knowledge of. There's so little we know – these days they're talking about other dimensions and so on. Could [these creatures] come from another place, and be slipping in and out of our [dimension] and perhaps that's why we can never find any physical evidence?"

After being assured that, as a yowie eyewitness, she was in good company – with hundreds of other people including rangers, police officers, surveyors, soldiers, and zoologists – Valerie laughed and said, "Well, now you can add a mental health worker to your list!"

Another Alpine Way encounter

John Lythollous was menaced by a somewhat hairier and more aggressive yowie on the same road, between Thredbo and Dead Horse Gap, only about 25 kilometres from the site of Valerie's encounter, in January 1997. (*The Yowie* p. 276)

Case 238. 2 Jan 2009. Oakview State Forest, near Kilkivan, QLD

Our friend Dean Harrison, founder of Australian Yowie Research, is one of a very few people who have come into physical contact with one of the creatures. His unforgettable experience occurred during an AYR expedition to a notorious yowie hot spot, Oakview State Forest.

> **Dean:** "At 1am I ventured away from camp, wandered down a granite cascade and sat next to a rock pool, with my Maglite and UHF radio beside me. The night was absolutely pitch black. The area was pine forest, to my left was a large hill.
>
> "Suddenly, something started marching along the peak of the hill. I radioed camp and, when told everyone was there, I said, 'Well, I have company!'
>
> "The moment I said that the creature turned and sprinted down, directly towards me – in total darkness. Its eyesight and agility were incredible – this is terrain I could not walk down without tripping over. It leapt over obstacles, jumped down onto a track above, then leapt again on the last descent, directly at me.
>
> "I was yelling for backup, and, with my hands out in front of me, yelling for it to stop, but knew it wasn't going to. It hit me in the middle of my chest with what felt like its forearm, and I landed on my back in the rock pool. POW! BOOM! I was seeing stars!
>
> "It had moved to my right, but I couldn't see exactly where it was, so I just rocked from side to side, kicking my legs in the air in every direction, trying to keep it off me, throwing futile punches, yelling and screaming for help. I was absolutely terrified.
>
> "Then, as the guys with their flashlights came down the cascade, it walked off into the bush behind me.

"Myself and a couple of the guys followed it. It kept its distance. Each time we stopped, it stopped and turned to look at us. Its eyes glowed a dull grey – you could see them blink. There was no moonlight or starlight to reflect – it was like they were self-illuminating. When it turned away, the eyes disappeared, but when it turned to look back, there they were.

"In the end, we gave up and went back to base camp."

The next morning, still severely rattled, Dean caught sight of a "black and heavy-set upper torso, shoulders and large, thick head" of another – or the same – yowie, which quickly dropped down out of sight.

"If there had been any other option/route to get back to camp I would have taken it, but I had to walk past it. I couldn't see it, laying there in the long grass, only about five metres away to my left – and frankly, I didn't *want* to see it. To say I was scared is an understatement. I had no idea what it was going to do, so I 'double-timed it' straight past.

"Then, only metres away to my right, came three mighty WHACK, WHACK WHACKs on a tree. I almost jumped out of my boots. Now I had one either side of me! From there, looking straight ahead, I did a half-walk, half-trundle, as quick as I could, trying not to seem as if I was running or scared."

Having spoken to him several times soon after the event, we can testify that, in addition to his physical injuries – very heavy bruising from neck to ankle – the attack left Dean, the most intrepid of field researchers, with debilitating mental turmoil for the better part of a year.

"It took," he told us, "about eight months to get a good night's sleep. I felt I had almost lost my life to these things for the second time. [The first time was in 1997, see *The Yowie*, pp. 85-86]. I now find it very hard to do field research as I used to. The fear factor remains and is overwhelming. The reality really set in – I feel that if I was alone that night, I would simply be yet another missing person."

Further sightings

Despite the nagging fear, Dean returned to Oakview State Forest, and, one afternoon in late 2009, in company with a retired NSW police officer, saw another yowie on a track about 100 metres from the site of the earlier incident.

That sighting was brief, and blessedly less scary – the dark, man-sized creature simply emerged from the bush, about 150 metres away, began walking towards them, noticed their presence, dropped to all fours, and hurried across the track into the scrub.

The policeman had glimpsed a similar creature while driving along the same stretch of track in August of the same year.

Case 239. 31 July 2009. Connollys Creek Gap, 5 km west of Mt George, NSW

Two days after encountering a yowie, Faye Burke and her cousin, Alana Garnett contacted the *Manning River Times*, which ran their story on 7 August. At around the same time, they were interviewed by yowie researcher Rex Gilroy and by journalists from NBN TV (Newcastle). Paul Cropper met them in Wingham on 30 August.

Faye Burke and Alana Garnett

On the night in question, they were towing a trailer from Wingham to Cundle Flat, on the Nowendoc Road. Life-long residents of the area, they knew the route very well.

> **Faye:** "We were going to pick up a load of pumpkins – I sell to greengrocers in Wingham, Taree, and Tuncurry. We thought, 'What a lovely night for a drive', because the stars were out – a sort of half-moonlit night.
>
> "There wasn't another car on the road, and I had my lights on high beam the entire way. That was really odd, because a lot of cars go through there. I think that's why it – he or she – tried to cross the road that night.
>
> "The clock on my car radio had just clicked over to 7.30; we were [climbing towards] Connollys Creek Gap, a winding road, going about 70 kph, when we

> came around a [left-hand curve] about 200 metres from the top, and my eyes caught this figure standing in the gutter on the left."

It was facing the embankment, with its back to them, and stood stock-still as Faye lifted her foot from the accelerator. "I slowed up, went around it, then put my foot down again."

The episode was very brief, but for a couple of seconds the creature's entire body was fully visible in the glare of their lights.

> "It had to be at least seven foot six to eight foot tall, and at least three foot six or four foot across the shoulders: just *massive*. Long hair on its head and shorter all over, well covered.
>
> "It stood erect and still, like it didn't want to be seen – like: 'Don't look at me – I'm just another lump of wood!'"

Its arms were tight against its body, but Faye thinks they would have extended down to knee level. "We never saw any ears, or feet – we just saw its back. Alana saw more than I did … we were only three feet from it."

> **Alana:** "Yes, I turned my head and followed it as we went past – I could have put my hand out and touched it – that's how close it was."

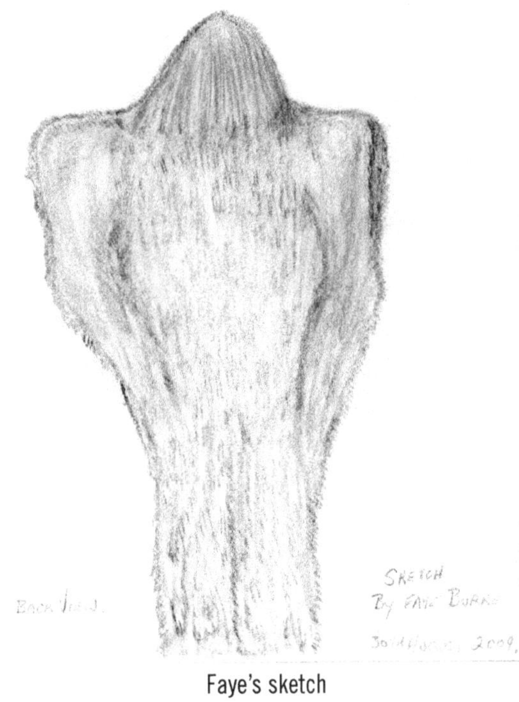

Faye's sketch

She, too, had the impression it was about eight feet tall and tremendously broad at the shoulder. Most of its "chocolatey-brown" body hair was about four-to-five inches long, except for dreadlock-like, matted patches on the shoulders. The head hair was quite long – "It tapered down [from the head to the shoulders, giving the impression that there was no neck]." As they "whizzed by" some of the longer hair "flicked up."

Chuckling, Alana recalls that as they accelerated away, Faye shouted several colourful words, eventually asking, "Did you see that?"

Faye says that Alana, "was spluttering like a child and hardly able to talk, but she said, 'It was a bigfoot!' I said, 'Do you mean a yowie?' and she said, 'Yeah – they're the same thing!'

> "I wanted to go back and take a photo with my phone [but] we were shaking … stunned … breathless [and] it was like, 'What if he attacks us? What if …?' Alana was saying, 'I don't want to see it again – no!'
>
> "We drove on, and about three kilometres from my brother's place I said, 'Alana – what if it jumped into the trailer, and we're taking it with us?' And she nearly died! [Laughs] And I scared myself by saying it! So, she shone the torch in the back and said, 'No – he's not, you silly bugger!' She still can't drive that road on her own in the dark."

On reaching Faye's brother's property, "We nearly fell out of the car, trying to both tell the story at the same time. He said, 'You're both *white* – something's happening here.' Every time we talked about it, truly, we got shivers up and down our spines – like it was happening all over again."

A footprint

Returning to Wingham in the morning, the cousins stopped and found a large footprint: "exactly where we saw him … the imprint of his toes were there in loose gravel – it was quite clear."

In the afternoon, they revisited the site with Faye's daughter Kim, who took several photographs.

> "As we walked around, we found scratches on trees [on both sides of] the road, like it had been looking for grubs or something. And there were tobacco bushes broken off, chewed, and thrown on the ground. Kim bit one and said it tasted like green apples. The bark was stripped off one tree, and about a metre away, there was a great big wet patch on the ground, and splashes on the leaves. I picked up a leaf, and I've never smelt urine like it. I was raised on a farm, I've smelt horses, cows, cats – but this just *stunk* so bad, and it was *oily*."

Because they were concerned that someone might be injured by the animal, the Faye and Alana contacted the *Manning River Times.* NBN TV soon got in touch, and the cousins guided a journalist and camera team out to the site – where something else of interest was soon noticed.

> **Faye:** "Near the tree with the stripped bark, a sapling had been pulled down and bent over. It hadn't been like that the previous day. I got goose bumps from my heels up to the top of my head when I saw that! The thing had come back!"

Yowie hunter visits

A couple of days later, veteran researcher Rex Gilroy went to the site with Faye and made a plaster cast of the yowie footprint, which he said measured 40 by 26 centimetres. Following a scent trail, he found a second, smaller footprint.

Although he said, "The footprints match ones I have found in the Blue Mountains", their shape – or at least, the shape of the larger one – adds an extra degree of strangeness to the whole episode. Kim's picture of the track, and Rex's cast, both appear to show that the creature was three-toed – unlike any known ape.

This is not the only instance of three-toed tracks being attributed to yowies. (e.g., Case 246 and *The Yowie*, pp. 83 and 108-109)

A psychic connection?

At one point during their conversation, Faye told Paul that she'd always known she had psychic ability and proceeded to relate a few remarkable episodes in which that ability had come to the fore.

As several other yowie eyewitnesses have made similar claims (e.g., Cases 235 and 259) Paul asked if she thought her psychic ability had any bearing on the yowie incident.

She was reluctant to make the connection: "No, maybe not ... well ... I don't really know." But she went on to say (twice) that "I was the chosen one to see him that night."

The cheerful, engaging lady concluded by saying, "I'm just so pleased there was someone else with me that night. I've never had a bad dream about it [but] Alana's had lots. I want to see him again. When I drive down that way now, I always throw out a couple of pumpkins for him!"

Case 240. 17 Oct 2009. Montville, Sunshine Coast, QLD

Jas Leontine reported this very unusual, very interesting, encounter to researcher Gary Opit on 19 March 2014.

> **Gary:** "Jas informed me that she and her partner Kurt observed what appeared to be a very elderly yowie at the Montville Lookout at 9.30 pm on 17 October 2009.
>
> "They had pulled into the lookout to view the lights of the Sunshine Coast far below and sat in their car in the darkness enjoying the view. When they restarted the vehicle and turned it towards the road, they saw, illuminated by the headlights, 15 metres away, a two-metre-tall, hunched, bipedal, ape-like animal.

"They observed it from behind as it walked off, entering thick vegetation. It had a bald head and its thin, muscular body, that looked very old, was covered with pale grey skin with veins clearly visible. The body was completely covered in sparse grey hair. The impression of the spinal vertebrae from the neck to the buttocks was visible.

"It did not appear to be a naked person and no normal human would be able to walk in complete darkness through vegetation adjacent a sheer cliff. Jas's mother had seen yowies many times as a child on a farm adjacent the Wollemi wilderness, but Jas had little knowledge of such things."

Case 241. 1 April 2010. 15 km north of Broken Hill, NSW

Martin Marinovich, a friend of the witness, Anton Margitich, contacted AYR on 12 April 2010. Paul Cropper and Tony Healy interviewed both men six days later.

By the time he was 37 years old, in 2010, Anton owned three businesses, including a parcel delivery company, Barrier Express. Twice a week he drove one of his trucks from Adelaide to Broken Hill and back. While at the Broken Hill end of the trip he stayed on a property owned by Martin, near Stephens Creek, about 20 kilometres north of town.

About 4 am on 1 April, another of Anton's mates (we didn't get his name) was driving him towards Martin's place in a four-wheel-drive that was equipped with "massive" lights because of the area's many kangaroos.

As they reached a creek crossing, the driver slowed to about 60 kmh. He was watching for 'roos on the right as Anton, in the passenger seat, was looking left.

Anton then noticed a big, hairy form right next to the track: "At first I thought it was a bear." It was looking down, but soon raised its head, and he saw it was more like an ape-man. As the vehicle passed within four feet of the creature, he got a pretty good look at it. "It looked angry – it looked straight at me."

It was at least a foot taller than his mate Martin, who is six foot four inches, and was covered with long, stringy, dark grey hair, except for its face, which was "a dusty-grey colour." Interestingly, because the creature's face was "weathered, wrinkled," Anton even ventured a guess as to its age: "It looked about 45 to 50 years old." The eyes were "a reflective, reddy colour – quite large – a bit bigger than a human's."

It had "massive hands", covered with long hair. He had the impression its feet were shaped like those of a human, and there was "long hair growing over them – like hair grows over the feet of some dogs."

The driver didn't see anything, but because of Anton's excited reaction, he also became quite freaked. He didn't want to turn around, so they continued on to Martin's property, five kilometres ahead.

When the two men arrived, Martin told us, "They were both extremely shaken, and it took a lot to calm them." He immediately told Anton to call the Broken Hill police, then grabbed a shot gun and a rifle and the three men returned to the creek crossing.

A policeman met them at the site, but the angry giant had apparently long gone. Because of recent floods, all the creek-side grass had been flattened, so they couldn't find footprints or other traces.

Case 242. June 2010. Otway Mountains, VIC, and April 2013, (exact location confidential) VIC

After contacting AYR, Glenn Souter was interviewed by Paul Cropper on 28 January 2014 and by Tony Healy on 10 March 2017.

These two events are of particular interest because few people are as familiar with Australian wildlife as Glenn Souter. A licensed breeder of native animals, he has spent thousands of hours in different parts of the country, in all weathers, day and night, observing and recording all manner of native and introduced animals.

As a veterinary nurse and lover of wildlife, his wife Soozanna is also very unlikely to misidentify any normal animal.

The 2010 yowie encounter occurred one Sunday afternoon, after he and Soozanna had visited friends in the Otway Mountains. At around 4 pm, before proceeding homewards, they drove to a nearby beauty spot, Beauchamp Falls.

Glenn: "It's a beautiful place – cool-temperate native forest and pine plantation. Between there and the Great Ocean Road it's all bush. No one else was there. There are tall tree ferns on either side of the narrow track to the falls, like a corridor. Soozanna was taking photos of mushrooms, and I was about 200 metres ahead.

"When I'm in the bush I like to walk very quietly, and I was either in the wrong place at the right time – or the right place at the *wrong* time. [Laughs]

"I heard a noise which I thought was a deer, a wombat, or a kangaroo. A cough of sorts – very low, deep, and powerful. So, I [stepped into the bush] about three or four

Glenn Souter

feet off the track, and something caught my eye about 10 or 12 metres ahead – and there, in amongst some saplings, was a large, browny-black figure.

"At first it was leaning forward and seemed to be looking at the ground. Your mind tries to make sense of what you are looking at – I so thought it was a large deer, browsing. Then it stood up and faced me – and my whole world changed right then. The hairs on my arms are standing up right now!

"This thing was immense, absolutely huge. It stood there looking at me for what seemed a lifetime. I was trying to take it all in and, basically, I was looking at a large, bipedal primate of sorts. It completely threw me – this thing was *huge* and looking straight at me.

"From that distance, I could see pretty much all of it down to about its ankles. The ground was level, there were small saplings, about as thick as broom handles, bracken fern and the odd tree fern, but basically, I had a clear line of sight.

"I was trying to take everything in [but] at the same time I was in complete and utter shock – my heart was probably racing at 100 miles a minute. We were looking at each other; I don't think it was sizing me up – I think it was more that it was shocked I was standing there, and I was shocked that it was opposite me. For the first time in my life, I was in the forest and felt like I didn't belong.

"The tree ferns were around 6 to 7 foot high, and this thing towered over them: it was *at least* eight feet tall ... extremely well built – there was muscle definition." He could see its chest moving as it breathed in and out.

"There is no doubt it was a real, flesh and blood animal. It was covered with hair, as far as I could see, from head to toe – a dark, chocolatey-brown colour. The hair on the head and shoulders was different from that on the rest of the body: it was longer, but from the shoulders down it was uniformly 3 to 4 inches long." There was no bare skin visible anywhere on the torso.

"I weigh about 67 kilos. This was over 200 kilos, without a doubt ... very, very muscular, and well built. It didn't look like a gorilla, and it didn't look like a man – it looked like a mismatch between the two. It had attributes of a human. I've been to the zoo numerous times with my children and stood in front of the glass wall where you can stand side-by-side with the gorillas – and you'd have to get two-and-a-bit gorillas on top of one another to come even close to [the size of] this thing.

"It was like the head was on right down on the shoulders. It wasn't domed, it was more rounded, as if you take an egg and cut it in half and sit it straight down [on the shoulders]. There was no visible neck: the trapezoid muscles and everything came straight up ... The shoulders extended at least two and a half feet on either side – giving it at least four-foot-wide shoulders – very square shoulders.

"Even though it was very solid and barrel-shaped, the body did taper down towards the hips. I wasn't able to see any genitalia or anything like that, but it was definitely a boy – I got that definite impression. The arms were very large and very long; they [extended to] around the knees." He didn't get a clear look at its hands.

"The eyes were incredibly large and deep-set. That's what really struck me – very large, dark eyes. I couldn't see any whites. The face was basically quite flat [but] the brow ridge was protruding. It had large, protruding lips – dark lips – quite a large mouth. Its nose wasn't too big; not a gorilla or orangutan-type nose. Definitely a human looking nose, but kind of squashed into the face.

"It didn't seem to have a beard, per se, but there was facial hair right up to just below the eyes. Basically, the only area that wasn't covered with hair was around the eyes and the nose [where] there was some dark skin showing.

"Its face was just rigid – a very blank expression – just intently looking at me. We stood looking at each other for probably 30 seconds to a minute – something like that. It then turned to its left and headed for the pine plantation.

"I gathered myself and was humming and hawing, wondering whether I should go back to the car or try to get a better look. So, I waited till it got out of sight and went back onto the track and ran up to where it had been standing and stepped back into the forest and got another good look at it from about 20 metres as it left the area.

"As it moved, pushing saplings out of the way, its arms never came up much past its waist [and he noticed] it had seven to eight-inch-long hair under the arms, from above the elbow to near the wrist. It was just brushing the saplings aside with no effort – just 'swimming' through the forest."

The yowie's gait wasn't quite human-like: "It was putting one foot in front of the other in a straight line – placing one foot dead in front of the other, and very long, powerful steps – it was covering a lot of ground with every step. He had quite a large bottom and massive shoulders. I could see the muscles moving in its legs.

"When it got to the verge of the fire trail [separating the native forest from the pines] it turned around, sort of pivoted with its hip – didn't use its neck – turned its body around, facing me again and gave me this look, as if to say, 'Don't even think about coming any closer!' That put the wind up me. There was this overwhelming sense of fear which I didn't have to begin with. I took three steps back, turned around and ran back up the track, and when I reached Soozanna I was as white as a ghost.

> "She hadn't seen anything. At that time, she was heavily pregnant and if she'd seen what I'd seen, she would have given birth right then and there! She knows all the Australian wildlife – different species and so on – but this would have completely thrown her.
>
> "It took me a couple of years to go back to the spot. I eventually went there to make a video, and all the time I was there my heart was racing and the hair was standing up on my arms. I couldn't get out of there fast enough! To this day I refuse to go back. Even in other forests now, I'm always looking around, checking things out. I used to be quite brave – I used to think I was the 'top bloke' in the bush, but now I know I'm not, and it's changed everything for me.
>
> "But that was the start of everything, really. It changed my whole life … I don't think I'll ever be the same person I was back then. That's what got me started in this whole yowie research business [laughs] and that led me to the last couple of sightings we've had."

Unknown to Glenn, there had been another sighting in the vicinity of Beauchamp Falls. In February 1984, a ranger, George Paras, encountered a five-foot-tall yowie just five kilometres to the east. (*The Yowie*, p. 254).

Second sighting. April 2013. Eastern Victoria (exact location confidential)

After the 2010 incident, Glenn, Soozanna, their mate Steve Crawford and others spent hundreds of hours searching for the hairy giants in other areas. After three years their efforts were rewarded when they came close to another yowie.

They began focussing on the area in question after a friend told them about a creature she'd seen there 20 years earlier.

Quite young at the time, she was out hunting with her father when they saw a large, furry animal they initially took to be a bear. They had plenty of time to observe it through a telescopic sight. It was on all fours, but finally stood up on its hind legs and walked away.

On their very first visit to the area, Glenn and Soozanna were fortunate enough to discover a set of large, human-looking footprints in an old logging coup at the base of a mountain.

> **Glenn:** "After the loggers burnt out the coup there was all this salty ash on the ground, mixed with sandy lime: a good place for tracks. I didn't have a tape measure or Plaster of Paris. I have a size 10 boot and the prints dwarfed my shoe – they were about 16-plus inches long and at least six or seven inches across.
>
> "They looked human … but not human … they looked a bit strange. There

> was probably four or five feet between steps, and they went right up, through Scotch thistle, into the scrub."

Toes – five of them – were clearly visible in the best-preserved footprint.

> "So, we left trail cameras up there and baited the area, but didn't have much luck for a long time."

By this time, Glen and Soozanna's family was growing rapidly – they had two-year-old twin boys and a newborn baby girl. Nevertheless, they continued to visit the site, night and day, in all weathers.

Finally, one dark night, as they were leaving the area to drive home, their luck changed.

> **Glenn:** "It was after 1 am ... headlights on high beam, and powerful spotlights on the bull bar that light up the whole road.
>
> "We'd gone around a few bends onto a straight stretch and there was a dip, and as I came up the other side there was this large, reddy-brown animal on the left-hand side, looking like it was getting ready to cross the road.
>
> "We both said, 'Yowie!' at the same time.
>
> "There was a large mountain gum beside the road and [the creature] was standing adjacent to that. We never got to see its face – it was side-on with more or less it's back to us. When we hit it with the lights it leant forward and bobbed down, and by that time I was driving past, so Soozanna got a better view than I did.
>
> "I was doing about 60 kilometres per hour; we turned around, came back, got out and ran to where it was standing. It was no longer there. [Then suddenly] there was a smell — like a rotten carcass – permeating from the forest. It came out of nowhere, then disappeared [even though] there was no breeze to make it waft. Then it came back again and that put the wind up us – and we jumped back in the car.
>
> "I had [the remains of a Kentucky Fried Chicken meal] in the car. I put the leftovers, including a container of potato and gravy with the lid still on, in a plastic bag, and left it where [the creature] had been standing.
>
> "The following night we found the bag at the bottom of the gully [next to the road]. Ripped open, everything gone. The potato and gravy container had had its lid taken off and there was a hole in the centre of the lid – as if something had pushed its finger through to remove it – and all the potato and gravy was gone."

Eye-shine

> "At 7:30 the next night we drove up there again. I had my Mag-Lite out the driver's side window, scoping the saplings and trees, and, right beside us, probably a few feet from the roadside, in amongst some acacia saplings, was a pair of blue-green eyes looking straight at us. We both got a good look at them.
>
> "It would go behind some trees and then put its head out again. There's no way it was a possum or an owl. At one stage, it put its hand in front of its face, because its eyes disappeared and then reappeared, and then there was a slow, deliberate blink.
>
> "There was no way I was going to get out of the car. I had my video camera, and I just didn't think about getting out."

Since then, Glenn, Soozanna, Steve Crawford and Glenn's father have returned to the site many times. On a couple of occasions, they've heard very large creatures crashing through the bush, and have had sticks and stones thrown at them from the scrub. On one occasion Glenn and his father were startled by "a very powerful scream, very primate-like – almost like something you'd hear at the zoo."

Black panthers

Yowies aren't the only cryptids Glenn and Soozanna have encountered – they've also sighted black panthers on two occasions – *while out looking for hairy ape-men.* Interestingly, at least four other researchers have experienced the same bizarre coincidence – sasquatch hunters Ray Crowe and Mike and Debbie Polesnek, plus yowie hunter Paul Compton.

Case 243. 11 Oct 2010. Springwood, NSW

This report was sent to us by our friend Neil Frost, whose own property at Springwood has been visited on-and-off by yowies since the 1980s. (*The Yowie*, pp. 89-117)

The witness, who wishes to remain anonymous, was interviewed by Paul Cropper on 29 Oct 2010.

The 50-year-old man had lived in the Blue Mountains for many years without seeing hide nor hair of the legendary yowie. On 11 October 2010, however, all that changed.

At 5.30 that morning, before beginning his long commute to Sydney, he drove to the local shops to return a DVD. Springwood is built on a few very narrow ridges, and parts of the shopping centre are just metres from the edge of the immense forest.

Stopping outside the video store on Hawkesbury Road, he turned his lights off, dropped the DVD into the box, got back in the car and reversed out of the parking space. As he turned his lights back on, he found, just six metres away, "this *thing* looking straight at me." Only the width of the street separated them. The creature was in the process of ascending a grassy slope from the edge of the bush.

It was at least six-and-a-half feet tall, like a large, very powerfully built man, but with "definitely an ape-like look." It was "brown, like a horse," hairy across the shoulders but mainly bare across the chest and stomach. The skin seemed to be a "chocolate-brown" tan – "like an Aborigine's skin colour."

The witness weighs 100 kg, but "this was easy 140-150 kilos." He had the impression it had slipped on the grassy slope just before the lights illuminated it. It was looking straight at the car and, because of the slope, was visible down to about its knees.

The head "seemed much smaller in proportion to its body size ... no perception of a neck – just head and shoulders." There was a "flat nose, but an emphasised brow and forehead." Its arms hung down to its knees. "It didn't seem aggressive ... I was more curious [than afraid]."

Although it was in the headlights for 20 seconds, he didn't notice any eye-shine – "just a glazed look; very much like a deer in the headlights. It didn't seem overly afraid," but looked "uncomfortable," so he switched off the lights, got out of the car and watched it walk away. There was sufficient light from streetlights to observe that it "just shuffled off, ambled off ... didn't seem to be in any hurry. The back was covered with hair, and it looked like it had a good head of hair." From behind, the head looked "almost chimp-like, without ears."

The episode may have lasted a minute to a minute and a half. "I was thinking [later] that I wish I'd filmed it – I had the I-Phone on the seat next to me."

Since the mid-1800s, the area around Springwood and neighbouring villages has produced scores of yowie reports. (See *The Yowie*, pp. 88-118.)

Case 244. November 2011. Wingello, NSW

Kayleigh (surname confidential) contacted AYR on 19 Jan 2012 and was interviewed by Paul Cropper four days later.

The township of Wingello is on the eastern side of the Southern Highlands, two kilometres from Wingello State Forest, which is bounded to the east by sheer cliffs that drop into the immense, yowie-haunted, Moreton National Park.

> **Kayleigh:** "I was driving home from Moss Vale on Penrose Road about ten o'clock at night. On the massive straight that comes into Wingello, there's a railway on the right, and open paddocks on the left. There's a row of trees on both sides.
>
> "I'd turned my headlights down because there was a car coming the other way, and I saw this thing on the left-hand side. It was facing away from me, so I didn't actually see its face. It was about six feet tall, at least, and furry.

"I was doing about 90 Ks an hour, and I had it in view for four or five seconds. It would have been better if I'd had my high beams on, but my lights were still pretty bright – I could definitely see it. [There was about two metres of grass between the road and the trees]. It was standing there, then it kind of slinked into the trees – took maybe two steps and was gone. It didn't seem startled – it looked like it just wanted to get away without being noticed.

"I saw it from about its waist to the top of its head. From the top of its head down it was covered in black or dark brown fur – maybe ten centimetres long, thick, not shiny, pretty smooth – not too roughed-up.

"But what really stood out for me was that it was really skinny – its shoulder blades stuck out as it walked. From shoulder to shoulder it was definitely not a metre, or anything like that; just like a solid-built man, but a bit taller. My boyfriend is about 90 kilos, so it would have been about 100 – not huge. Its arms weren't massive, but they were long … I didn't see its hands.

"I went, 'Oh gosh!' I was so scared – it was the scariest thing I'd seen in my life. I looked back and thought, 'I should have stopped,' but I was petrified. My mouth literally dropped open, and that's never happened before. I didn't know what to think – and then I remembered my friend Sean [who'd recently seen the same, or a very similar, animal] and I thought, 'My God – that's it!'

"He saw it about six months earlier. He was driving [from Canyonleigh] to Moss Vale at 6.30 in the morning. He got a pretty good look – it was on the side of the road, and when he got close it ran into a paddock. He said it was a big, lanky, black thing – and very quick on its feet."

As in several of our other yowie witness interviews, another Australian cryptid, the mysterious black panther, crept into Kayleigh's story:

"Oh yes – there's lots of [panther] stories around here, down in the gorge and all over the place! My [twin] sister's seen one. Cattle have gone missing. Friends of mine at Tallong had one of their cattle dragged up a tree. We all believe in the panthers around here. There were a lot of stories about three years ago."

Case 245. Summer, 2011. Near Donnybrook, WA

The witness's wife contacted AYR on 18 March 2014 and Paul Cropper interviewed the witness on 3 April.

This report is unusual because the eyewitness, "K" is Canadian – a Native American from the Great Lakes area.

Western Australian yowie reports are relatively rare, so it's interesting to note that the sighting occurred only 25 kilometres south of the Collie River – site of Michelle Netos's 1991 encounter. (Case 216)

> **"K":** "It was about three in the afternoon. I was building some chicken coops. I'd been up to the tip to get some wire and was driving back down – it's a very winding and steep road. The bush is not real dense – a lot of [the underbrush] had been burnt out, so you could see a fair ways through it."

When he first noticed the creature, he assumed it was a tree stump, and concentrated instead on a truck coming from the other direction.

> "Then I looked again, and it looked like an orangutan [and] when I slowed down for this truck, that's when it took off.
>
> "And this fellow who was driving up the hill … we both seen it at the same time, because when I stopped, he stopped, and he was looking at it run down the hill too.
>
> "It was kind of running and hopping – it hooked onto a tree and [swung] around to another tree and [swung] around that, using the trees to propel itself – to get going quicker. He kind of flew through the air. He'd grab hold of a tree and fly through the air a fair ways before he landed again. He was moving *fast*. It's a very steep slope and he was really flying down!"

At the closest point, the yowie was about 40 metres from "K", and the episode lasted no more than 15 seconds. "He was down the hill very quick, and into the bush. I was in pretty good shape at the time, but I couldn't have gone down the hill that quick.

> "I'm five foot ten, and it was a little bit shorter, but a hell of a lot stockier. I weigh about 115 kilos; this guy would have went about 150. You could see the muscles on him. His arms weren't really long like an orangutan's, and he didn't move like a monkey or orangutan or gorilla … it ran like a man.
>
> "Around the shoulders and arms [and head] the hair looked fairly long. It was a dark reddish, not really red like an orangutan – more of a browny-red, kind of matted – he looked kinda rough looking.
>
> "I didn't get a good look at the face – you're in sort of shock – you don't expect to see a big monkey running in the bush! But from what I could see, it didn't have a kind of protruding face like an orangutan would.
>
> "This other fellow, I wish I could have got to chat with him because he seen it

too, and he had a look of shock on his face as well. We were stopped right across from each other, and we just kind of looked at each other, like, 'What the hell was *that*?' But we were both kind of stunned, and he just drove on – and I drove on in the opposite direction."

Being a Native American, "K" was familiar with the sasquatch phenomenon. "Yeah – I'm very familiar with that, back home". As for the yowie, however, "… it was the farthest thing from my mind – never thought I'd see anything like that, ever. Back home, when I was around Jasper, Alberta, and places like that, I thought I might get lucky enough to see something, but I never did.

"I only ever told my wife and a friend, Big John [about the yowie incident]. I didn't think anyone would believe me. I looked around town for the other fellow's ute, and if I'd seen it, I would have talked to him about what we saw. It's not a real big town, but I didn't see him around."

Case 246. 2011 – 2012. Near Deer Vale, 16 km west of Dorrigo, NSW

This remarkable episode was thoroughly investigated by naturalist and broadcaster Gary Opit, between May and November 2016.

Gary's involvement began when he received a series of emails from the principal witness, Ben Tracy, between 4 and 19 May 2016. With Ben's permission, Gary lightly edited those emails and combined them to produce an excellent record of the events.

The following was compiled from Gary's records and from an interview Paul Cropper later conducted with Ben.

Ben: "I had several encounters with what I would describe as 'Gorilla Men' [while] I was a member of a small fencing crew on a cattle property, "Coolahwarrah," 1,208m above sea level at Coutts Water, west of Deer Vale.

"We installed 26 kilometres of fencing over nine months during 2011 and 2012 [and] lived in a large, corrugated iron shed, which had an open garage. Within the garage was a doorway to the living quarters – three bedrooms, kitchen, and laundry. The property manager lived in the main house 60 metres away. Both buildings were situated on a large, treeless, level paddock.

"[After work] we used to visit a waterfall for a swim, and one time I saw a large animal, like an ape on all fours, looking at me from the top of the waterfall

... it quickly pulled itself back and disappeared. I was the only one that saw it, primarily just the head, and did not know what it was.

"Gorilla Man Falls". (Gary Opit)

"We had our first clear sighting of the 'Gorilla Man' at sunrise [5:45am] as we drove over a hill to work. We saw the back of a large black animal on the fence line 100 to 120 metres away, and at first thought it was a black cow. Just as we saw it, it saw us, stood up and turned its upper body, turning at its hips, to look at us, while its feet remained facing away.

"It looked like an upright gorilla, though it did not have a big gorilla head and the body was covered in short black hair, which looked shiny in the early morning sunlight. It was between 2.5 to 3 m tall [about 8 to 10 feet]. Its shoulders were 1.5 metres across, with long arms, with huge, muscular upper arms. It looked top-heavy, with a skinny midriff compared to the width of the shoulders.

"It then ran off on two legs, extremely fast, for 150 metres across the open paddock, so that we had it in view for a couple of minutes. It was running in a manner quite impossible for a person, because it ran with its body leaning far forward and crouched down, so that for a moment it looked like it was running on all four limbs like an ape.

"Then we observed that its forearms were very rapidly swinging very far forward and very far backwards, so that at the top of the swing the arms were horizontal with the head and shoulders, like a pair of pendulums. The clenched hands were visible above and level with the top of the grass as they swung down. The grass was around half a metre high, and its knees were only just protruding above the top of the grass.

"It wasn't human" – Ben Tracy's sketch.

"The foreman and crew were stunned and frightened, but as it had run away and we had a job to do, we just got back to work. [When first sighted, the creature had been crouching beside] the fence line where we had driven in steel pickets the evening before. It had pulled 10 or so out and had twisted and bent a few.

"We often discussed what we had seen [and] agreed it wasn't human or anything we had seen before. When we finished building the fence, we found a dead possum strung along the barb wire, which had been purposely placed there, with a slit along the belly with the stomach and intestines removed. It appeared to us to be a peace offering or gift from the 'Gorilla Man.' We also often found hair on the fence."

Other strange events

"There was a tree on the riverbank that we used to hang our towels on when we swam. One day we were surprised to find the tree was gone … snapped off four inches above the base.

"On the side of the waterfall was a natural basin that received water only when the river was in flood. It was about 750 mm deep and 1.5 m wide, like a spa bath, full of greasy, stagnant, filthy water with hair floating in it. I found a skull on the side of the basin, flat-faced, no snout, round like a small ape, with a ridgeline along the top. The teeth closely resembled human teeth except for the canines, which were large, round, stout, and slightly longer than human canines. It was perfectly clean and looked like it had been lying there for some years. I thought that the skull looked like a baby 'Gorilla Man' and that they may have placed the body of their child in the pool as a water burial.

"The foreman panicked when he saw me handling it and told me to throw it back in the pond and he and the other crew member began to run back to the car. I was concerned that the 'Gorilla Man' may become enraged by me handling its baby's remains, so I threw the skull back into the pond and ran.

"On a cliff face about 2.5 metres above the river was a perfect, round little cave or rock hole 25 cm in diameter. I climbed up to see what was inside but saw nothing. Next time we were down at the river, there was a rock pushed into the cave, plugging it up – as if the 'Gorilla Man' had observed my actions and was upset at me looking into it. Upstream of the big waterfall was a smaller waterfall and beside it was a cave, the rocky floor of which was covered with fresh green ferns and leafy branches, where something had been sleeping."

More sightings

"Our next encounter was on a weekend, around midday, when we went down the path beside the waterfall … [they heard] a cracking sound, like claws scratching and breaking off pieces of wood behind a huge, rotten, fallen tree – sounding like an animal searching for grubs.

"I was in front, halfway across the creek when I heard something extremely large and powerful running away from the dead tree right in front of me, with the trees and shrubs shaking as it passed, moving up the slope at great speed. I turned to ask my mates if they had seen it and … saw one workmate trying to claw his way back up the embankment in terrible panic. The foreman was running back towards the waterfall with another workmate running close behind.

"They said later that they had seen the 'Gorilla Man' and one of my workmates shit his pants in transit. He left the job that day, never to return. I didn't get a look at it, but the men [who did] were clearly shaken. When we returned to the car, we found that the 'Gorilla Man', or another one like it, had … opened the doors. We had left the doors closed, but not locked.

"Gorilla Man Falls"

"On the weekends we used to go brown trout fishing and once, when I was with a workmate, we saw another one of the 'Gorilla Men'. It was around half a metre shorter than the one we saw at the fence – about 1.8 to 2 metres tall – and it was running in an upright position, almost like it was jogging, 80 metres away towards the forest. Like the first one, it looked top heavy, with a skinny midriff compared to the width of the shoulders, though its hair was longer and shaggy."

Night stalkers

"We [also] experienced unusual things at night. On several mornings we found that someone had opened the car doors while we were sleeping. One night a 'Gorilla Man' came to our shed and harassed us – banging on the walls, rattling the door, and bouncing the front end of the Toyota Pajero 4x4, nearly shearing the bolts on the bull bar.

"It was the most terrifying night of my life. [The creature] was less than 3 metres from me with only corrugated tin and a huge plate glass window between us. I pulled my swag flap over my head, held my breath, played dead and passed out. One of the crew, who had his bed in the laundry, [was] so frightened that he had covered himself and his bed with all of the dirty clothes piled beside the washing machine, hoping that if the 'Gorilla Man' entered the house, it would not notice him.

"In the morning, we found dozens of footprints, each about 35 cm in length or more, in the fine bulldust [on the concrete floor of the garage] leading up to the door of [our sleeping] quarters and around the car. Each footprint had only three toes, and the dermal ridges of the toes and the foot were clearly visible.

"One night we got bored and [went] bush-bashing for fun, through the cleared paddocks. The foreman drove the car, with me beside him and Steve in the back seat. Suddenly Steve began screaming and yelling that the 'Gorilla Man' was running right beside the car! The foreman hit the brakes and we felt a jolt and thud! I opened my door to get out and have a look, but Steve pleaded with me to stay in the car. There was no sign of the 'Gorilla Man' [but] there was a large dent on the rear left quarter panel.

"Steve became a changed man, much more paranoid … no longer self-assured … so terrified that he packed up and left the next day. He phoned his mother to come and pick him up and waited at the front gate for three hours for her to arrive, refusing to re-enter the property.

"When the job was finished, we went to Armidale and stayed [there] overnight. [That evening] the property manager phoned the foreman on his mobile phone. Thinking that we were still at the property, he told us to stop throwing rocks onto his roof – the main house, 60 m from the shed. He refused to believe there were no workmen on the property, as he knew nothing about the 'Gorilla Man' and believed that only a person could be throwing the rocks.

"I am now obsessing about this and need closure. If you know anyone to accompany me to the wash pool to retrieve the skull, please contact me, as my workmates want nothing to do with it. The 'Gorilla Man' was skittish but inquisitive. I don't think it was being too aggressive; it clearly could have had us if it wanted, in many situations."

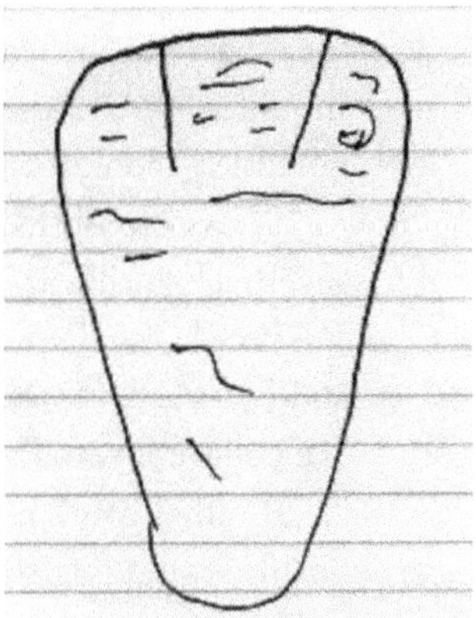

"Gorilla Man" footprint. Sketch by Ben Tracy

Responding to Ben's appeal, Gary Opit visited "Coolawarrah" with him on two occasions. During the first expedition, on 21 May 2016, Gary and his mate Henning Beth joined Ben and his Aunty Helen for an overnight camp within two kilometres of the "Gorilla Man" site. After a quick chat to the wife of the new station manager, they hiked to the falls.

> **Gary:** "The creek was flowing strongly, and we first encountered the smaller waterfall with the adjacent cave. We named this Cave Falls. We then followed the river downstream and reached the larger waterfall which we named Gorilla Man Falls.
>
> "After resting on the top of the falls we descended and Ben examined the pool, searching for the skull, but found nothing."

The "Gorilla Man" episode was a truly remarkable case, thoroughly documented by naturalist Gary Opit and well-illustrated by his photographs and Ben's sketches.

Case 247. 24 Apr 2012. 25 km east of Miriam Vale, QLD

At around midnight, while driving home after completing two weeks' work in another area, "C. R." and his mate encountered a yowie on Tableland Road, just south of the intersection with Fingerboard Road. Shortly after the event, he sent the following account to AYR:

> "After the second floodway crossing, we seen what can only be described as a large primate-like creature on the right-hand side, where the scrub was cleared. It would have been [no more than] 10 metres from the car … hunched over something, [maybe] some roadkill, with its back facing us.
>
> "Even while squatting, it would have been around five feet tall – at least the height of the [4WD] ute we were travelling in – making it at least 6.5 feet standing upright. It did not even react to the car, unless it did after we'd passed.
>
> "Both of us let out a big, 'What the … was that?'
>
> "Broad shoulders … very little neck – just a large head, thick, matted hair – it must have [been] reasonably long. The body was solid, a straight figure from the shoulders down, the shoulders looked muscular due to their width, but I was not able to look for long enough to tell if the whole body was muscular.
>
> "I have stood, face to face, with a piece of glass separating me from a male silverback gorilla at Taronga Park Zoo, but this creature looked much larger.
>
> "People ask, 'Were you on drugs? How tired were you?' The answer is 'No'– and I was fully awake. My mate and I were stunned [and] agreed on every detail. We believe it was a yowie.
>
> "If only I could take the image from my mind and replay it to you and others, people would realise what we had seen.
>
> "It's funny – the old saying: 'Don't believe it till you see it.'"

There have been at least two other sightings in same area. On 26 October 2018, a Dutch backpacker, Stef van Wijchen, saw a Wookie-like creature about six kilometres to the north (Case 268) and, on 4 May 2014, Angel Owen saw an eight-foot-tall ape-man in Deepwater National Park, about 15 kilometres to the north-east. (Cases 248, below)

Case 248. 4 May 2014. Deepwater National Park, QLD

Angel Owen contacted AYR on 10 May 2014 and was interviewed by Paul Cropper on the same day.

> **Angel:** "Our property backs onto the national park. I've lived here all my life, and my parents have lived here for 24 years.
>
> "There's always weird stuff happening. About two years ago we were woken up by a screaming noise – my dad thought it was someone in the bush screaming 'Help, Help, Help!' So, he rushed out with a torch. The torch batteries suddenly went flat and then he saw two massive red eyes, about nine feet off the ground.
>
> "Dad is about six foot two, and he started throwing rocks at it, and it came towards him. Then he saw another one on the other side coming towards him, and he ran back and locked the house up and wouldn't go back out there for days.
>
> "A few days later, it happened again, and it was banging on the roof and running around the back of the house for quite a while. It cast a massive shadow."

Angel and the "devil-devil"

> "[On 4 May 2014] me and my dad set out about 9 am for a motorbike ride through the national park. We went about five kilometres down the track and came to a swampy area … so we turned around and started back.
>
> "Dad was ahead of me when my motor bike stalled near a fire trail … I saw something in the tree line, like a shadow; it looked like it was crouching down. I was straining to see what it was, and it stood up and walked into the middle of the track.
>
> "It was taller than the tree regrowth and they were about eight feet tall. It was really broad across the shoulders and looked like its head was just sitting on its shoulders. It was hunched a little and the arms hung down just below the knees. It was dark brown, almost black. There was no way it was human.

"It's chest, arms and legs were huge. It had its chest out – like it was trying to intimidate. It was like nothing I've ever seen before. It didn't move – just stood there looking at me.

"I had this feeling of danger – it was scary. I took off after my dad and was yelling out for him, but he couldn't hear me until I got onto the top track ... I told him what I saw, and we rode home.

"For the next week, I wouldn't even walk to the bus stop alone."

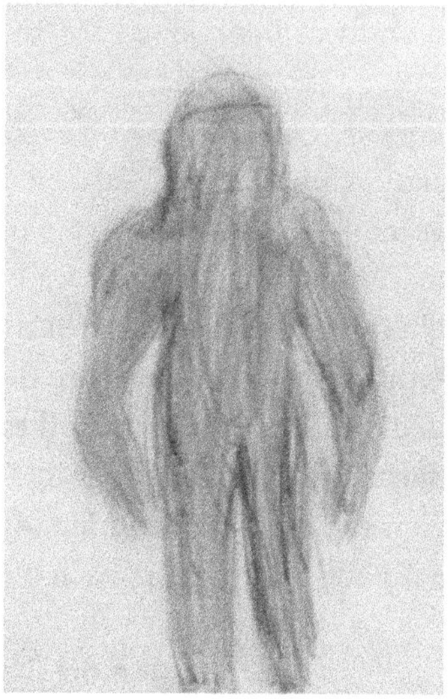

Sketch by Angel Owen

Angel's second sketch

Angel was a very compelling witness. Two things she mentioned might be of particular significance: her father's torch batteries suddenly going flat just before he encountered the two yowies outside the house, and her motorbike engine stalling just before she saw the creature in the national park. Electrical malfunctions and stalled engines have been mentioned in other yowie (and sasquatch) reports.

Case 249. 6 October 2014. Wilpena Pound, SA

After contacting AYR, Dylan Grantham was interviewed by Paul Cropper in December 2014.

> **Dylan:** "The country is all dry, bushy land; rock caves up there, plenty of water [because of] natural springs. It's about 30 minutes away from the nearest town – that's Hawker.
>
> "It was [just after 3 pm] on a sunny day. I was up there looking for reptiles, and I was at the creek, down on my side, filling my canteen.
>
> "And I looked [upstream] and seen this big, black figure – a black, ape kind of thing – about 40 to 50 metres away. It was in the creek bed … behind a tree, but its head was out a bit, watching me.
>
> "It was standing straight up, and it looked down at me for a couple of seconds and bolted off. It moved real quick – one, two steps, and then *gone*.
>
> "I was a bit shocked. I was thinking, 'Hang on – this is something not very human!' It was a very ugly looking.
>
> "It didn't make a sound … It was seven to eight-foot tall … hairy … the shoulders were really bulky – solid. The arms were long. It wasn't a guy in a suit, because it had hair down its [ape-like] face. The only place where I could see he had no hair was the front part of his hands. The hair was black, black to brown. Long, not smooth – like, wild.
>
> "I weigh about 71 kilos. It'd be about 250 to 300 – four or five times the size of me. Solid – I've never seen shoulders like it before. And its legs were pretty friggin' big too.
>
> "It didn't run like no person – it ran taking really giant steps and it was waving its arms … It moved like an animal, like a monkey – it was like it was jumping.
>
> "Before I went to the creek, I had a feeling something was watching me … I ignored it.
>
> "I've been looking for reptiles all my life and I've never … I've heard stories [about yowies] from the Aboriginal people but I'd never believed them until that day. Up there it is really sacred Aboriginal land and I know a few Aboriginals, so I'm allowed to go on it. Not many people go there.
>
> "I've been up there camping before and have heard things like … banging into the trees and strange noises … like a stick being banged onto a tree.
>
> "I went back the next day and didn't see nothing. I got a little bit of hair, though, from where the tree was. It was hanging on the tree, on a branch. It's about 20 centimetres long."

Possibly because it's a large, arid state with a very small population, we receive few yowie reports from South Australia. We do have, however, another good modern-era report from Roxby Downs, about 200 kilometres north-west of Wilpena Pound, on the western side of Lake Torrens. (Case 221)

Interestingly, during an 1851 expedition, explorers were warned, by Aborigines to the west of Lake Torrens, about savage, ape-like *jimbras*. (Case 23)

Case 250. 2015. Manumbar, 25 km south of Kilkivan, QLD

Rod Benfield contacted AYR on 20 Oct 2019 and was interviewed by Dean Harrison the same day.

The incident occurred on Manumbar Road, between Wrattens Forest and Gallangowan State Forest.

> **Rod:** "I used to go out to the back of Manumbar to get logs ... with my brother-in-law, in his nine-ton truck. It's a dirt road for a fair way.
>
> "Around 9 o'clock in the morning, we were heading down the road when I needed to go for a pee. [His brother-in-law] pulled over to the side of the road. I got out and for some reason I decided to go through the fence. I crawled through the barbed wire and headed towards a tree [that was] about six-foot across.
>
> "It was flat ground, mainly gum trees and brownish ... grass, and behind the trees was a creek. I lent up against the tree to have a pee, looked up, and this very large, cranky, man-like gorilla-type thing was hunched over, looking like he was going to launch. He had one hand on the ground like a football player, ready to launch off the ground. He was about 10 to 15 metres away and I was about 10 metres from the fence.
>
> "He came charging straight at me ... ripping anything out of the ground that got in his way [and made] ... I'm not sure if you would call it a growl, it was more of a roar.
>
> "I turned and ran. Funny thing was, I [got] snagged on the fence on the way in, but I didn't even touch the barbed wire on the way out – I slid under that bottom wire and it never touched a piece of clothing. I slid all the way back to the truck. I was moving!
>
> "He stopped at the tree I'd been at. It was open ground from the tree to the fence. Thinking about it now, I think it was a mock charge. A 'Get out of here, I don't want you here!' If he'd really wanted to hurt me, he could have run me down in two seconds

"This gorilla-type thing … hunched over."

"He came charging straight at me."

"He was really tall … crouching down, he [was] easily five-and-a-half feet at the shoulders. Standing up … about seven-and-a-half to eight feet. Big [long] arms … big chest … big hands – he was a big boy – extremely athletic looking. Any bloke would be proud to have a set of shoulders like that. The neck was a little bit shorter than a normal neck ... a thick neck with big trapezius, which I guess is why people think they have very short necks.

"[It wasn't] as hairy as what some people say – I could see a lot of muscle ... see through the red/brown hair to the skin ... his chest hardly had any hair – dark, tanned skin that looked like leather.

"There wasn't a lot of hair on the face. It looked like a gorilla [but had] some human features ... a normal type of nose, but big, broad, flared nostrils. Under the eye ridge he had big, sunken eye sockets ... the eyes were shaped like ours but a bit bigger, with a bigger pupil and he had the whites of his eyes. His eyes opened right up; he certainly didn't like me. He had big chiselled-looking teeth. Gorillas have really big canines [but] he didn't. I didn't look at his face for long, it wasn't a pretty thing – there was a bit of drool going on, he was pissed off, he was angry.

"If I was to estimate an age, I would say maybe early thirties He was fairly well weathered. The skin was a bit like a gorilla's skin, leathery looking, like it had been out in the sun a long time.

"I got into the truck and said, 'Did you bloody see *that*?' And the brother-in-law said, 'Did a deer chase you or something?'

"I don't know where [the creature] went, he was just gone. There was still some dust in the air. I don't know if he went behind the tree or shot back down to the creek.

"Afterwards, it probably took me a good 18 months to go into the bush without my brother-in-law or my wife. That type of thing doesn't go away. I still won't go far into the bush by myself. Even going in on a horse while mustering, I'll wait for everyone else to go in first.

"If they want to catch you, they're going to get you. You won't get away from them. With the amount of power they can unleash, what chance have we got?"

Case 251. 2015 and 2019. Gympie, QLD and Mansfield area, VIC

Tenia Kyriaki contacted AYR on 15 April 2021 and was interviewed by Sarah Bignell on 16 May 2021.

Her report, as will soon become apparent, contains several "high strangeness" elements. That's not to say, however, that we disbelieve her – she impressed us as being entirely honest and sane.

First sighting. QLD 2015

The incident occurred while Tenia, a cheerful, confident lady who was born and raised in Greece, was visiting a friend on a property off Thomas Road, Curra, about 12 kilometres north-west of Gympie.

Curra is a rural residential ("hobby farm") locality, with a population of 1,920. Most dwellings appear to be situated on six to ten-hectare blocks and most either face onto or are surrounded by bush. It is certainly not a wilderness area, but Curra State Forest is only three kilometres to the east.

> **Tenia:** "It feels so good to be able to tell people about it without being regarded as crazy.
>
> "The house is secluded, on a small slope [with a 20-metre clearing around it]. It was morning, I was standing on the porch looking downwards into the forest and this large, reddy, orangutan-like creature pushed the trees aside and just walked out, waving its arms up and down like it was very happy to see me. It had a grin on its face, and it started walking towards me, waving – you know how children [wave] their arms up and down? Like, 'Oh – there's my friend! Let's go for a walk,' or something. I could see eagerness on its face.
>
> "I just stood there, looking at it. I couldn't talk – because you're trying to work out what you're looking at. I didn't know about yowies. All I could think of was an orangutan had escaped from a zoo, or someone had been keeping animals illegally, and I said to my friend, who was sitting behind me, 'Have your neighbours got monkeys?' He got up to see what I was looking at [and] he went white, his jaw dropped … he was ready to faint!
>
> "It [had been] walking towards me until my friend stood up, and then it stopped and put its arms down – you know how we do when we're disappointed. He [looked] very sad [but] didn't turn around. And my friend took me inside and went around locking the doors and windows.
>
> "I would have allowed him to come closer. I reckon he would have walked right up to me, but when he saw my friend, he sort of lost his bearings, like, 'Oh – there's someone else there!'
>
> "I was looking outside the rest of the day; I could feel something looming outside and so could my friend. It was really strange, we looked at each other and didn't talk about it, but he said, 'Please don't walk into the bush!' I could see he was frightened. I only stayed there a few days and he said, 'Oh God, how am I going to stay here alone now?' Men are such chickens!" [laughs]
>
> **Sarah:** "How tall do you think the creature was?"
>
> **Tenia:** "It was *big* – seven foot; a big being – I don't want to call it an animal.

It wasn't really an ape; it's face … had more human features. [The head] had a little bit of a cone on top – not like a human head. Not much of a neck."

Sarah: "You said it had a 'grin' on its face?"

Tenia: "Yes! The mouth – it had a grin [but] I can't remember the teeth. And the eyes were open, like, 'Hi there – let's go for a walk!' Like a person.

"The face was black, and the chest was black, and when it lifted its arms, I could see hair hanging from under them. Very, very long arms – when they were going up and down, they extended [very widely] and … very large hands. Very broad shoulders – three feet across. The legs were bow, not straight. Gingery-red hair."

Sarah: "Was hair covering the hands?"

Tenia: "On top of the hands, yes, but when he lifted his hands, I saw the palms were black."

Sarah: "Did you get the impression it was male?"

Tenia: "I think it was a 'him.' First of all, the chest … the male form."

Sarah: "Could you estimate its age?"

Tenia: "I think it was young – the movement and everything. I could see the trees moving behind him – maybe there were others, family members, back there, or he was the smallest one. I don't know."

Sarah: "How long was it in view for?"

Tenia: "I don't know, it could have been ten minutes, it seemed like an eternity."

Sarah: "Did you notice any noise? Vocalisation?"

Tenia: "No, I couldn't hear anything. You know how time stands still? He was in front of me, and you know when you watch a movie, and they stop the sound? That's how it felt – it was surreal. That image was *surreal.*

"It was something that appeared for me – it was like he knew me. I consider myself very special to have seen that."

Sarah: "Are you a spiritual person … psychic, intuitive?"

Tenia: "Yes – I realised I was looking at something special that other people don't get to see. That sounds very egotistical, now that I say it, but I thought, 'Wow, God – what have you presented to me here?' It was the most beautiful thing – absolutely beautiful – the image will stay with me forever.

"In all these stories [that she has since read about yowies] it seems like they are seen by chance, but it walked right out in front of me – just *presented* itself to me. I'm honoured to have seen it, and I reckon if my friend wasn't there, it would have walked right up to me."

Second sighting, Victoria, 2019

Tenia: "This happened at 8.30 in the morning, near Mansfield, on the road to Benalla. I have a student up there. It's a long drive, so I like to get out of the car with my coffee and stretch a little. As I was leaning on the car [she noticed] a strong smell, sort of a toxic … sulphur … not overpowering. [Then a branch suddenly landed on the car] and I turned around.

"This one was a very tall, black, scarier version. A huge thing – eight feet tall. It was close – like, 5 to 6 metres away, hiding behind a tree, holding the tree. I could see his head and hand and a bit of shoulder. I'm not very tall, so as I looked up, I saw the black, black eyes. His face was very black, everything was black. The hair was dark; I could see no white in the eyes.

"I don't know if it was a male or a girl, because I only saw the head and shoulder – but it was a huge creature – a King Kong! [Laughs]

"I just ran, got in the car, and bolted! [But] I'd had my back to him, I'm a small, little woman, it would have taken only three strides to reach me, if he'd wanted to hurt me. And the way he threw that branch, it was a soft throw to get my attention, like: 'Hey, I'm here!'

"The first one was playful; this one was shy – scared or shy. It [seemed] quite young. Very groomed, the hair all shiny, like someone had just combed his hair.

"Some of them might be dangerous, but I don't think they intend to hurt people. I think they're very child-like, even though they're huge. They're beautiful, and I think we should protect them.

"I don't think they'd present themselves to a group of people, because they'd feel threatened. They're nature-loving creatures and that's where they belong – and we are destroyers."

While the many "high strangeness" elements in this story might trigger "high scepticism" in some readers, they don't affect us that way, because most of the weird details – the feeling of time standing still, the uncanny silence etc., – have been mentioned before, in obscure reports that Tenia could not have been aware of.

Her belief that the "beautiful" first creature "*presented* itself" echoes what Margaret McRae and Glenys Devine said of the creatures they encountered at Jiggi Valley in 2003 and near Lithgow in 2004. (Cases 228 and 229)

Both of those women said the "beautiful" male creatures appeared to be quite young. Glenys' creature was "showing off"; Margaret's was "showing himself to us."

Tenia is also not alone in claiming to have encountered yowies on more than one occasion in widely separated areas. Nor is she the only "repeater" to claim psychic ability. (e.g., Cases 235 and 259)

There is, however, one detail in her story that is unique, at least in our experience – the smaller yowie's friendly "grin." In Chapter 6 of *The Yowie*, after listing the various facial expressions witnesses have described throughout the years – “poker-faced”, “grimacing”, “shocked”, “sad”, “frightened”, “angry”, etc., – we jokingly wrote that “We're still waiting for someone to report a smiling yowie.”

Makes you wonder what's next – singing, dancing ape-men?

Sarah Bignell

Sarah Bignell, of Castlemaine, Victoria, has been researching the worldwide bigfoot phenomenon for over 16 years.

On her podcast, Yowie Central, launched in September 2019, she not only interviews many yowie eye-witnesses, but also takes listeners on a deep dive into all manner of other strange phenomena, including out-of-place big cat sightings, encounters with ghosts, UFOs, Min Mins and ultra-weird creatures such as “dogman.”

Her high levels of empathy and respect enable her interview subjects to relax and relate their often quite remarkable experiences in astounding detail.

Now a major figure in Australian cryptozoological research, Sarah is a key member of the AYR team, working closely with Dean Harrison and his associates, conducting interviews and delivering regular updates of their expeditions and discoveries.

Yowie Central can be found on all the major podcast platforms. To report a sighting, contact Sarah at yowiecentral@gmail.com

Case 252. 10 Mar 2015. Near Canunga Land Warfare Centre, QLD

Robert Cook contacted AYR on 26 Aug 2016, and was interviewed by Paul Cropper two days later.

The incident occurred between Beechmont and Canungra, on the western boundary of the vast Canungra Land Warfare Centre (CLWC).

> **Robert:** "I was heading down the mountain on Beechmont Road, travelling at about 60 kilometres an hour. It was daylight, 4.30 to 5 pm.
>
> "As I came around an S-bend, it came from the right-hand [CLWC] side, about 20 to 30 metres ahead. I was sitting up high in my truck and had a very good look at it. It was huge! Its head was level with where I was sitting ... friggin' massive – eight to ten feet tall.
>
> "A big barrel of a body. Shaggy, reddish-brown hair all over; long arms. No neck – head right down on its shoulders. I didn't see a face.
>
> "It was the body mass that got me – how *thick* the body was. *Humongous*!
>
> "There was no friggin' way it was a guy in a ghillie suit. It walked across the road in *two steps!* I know what a big person looks like, but I don't know anyone who can cross a road in two steps!
>
> "Its legs were long, but it looked like its knees were really low ... it had a smooth walk, but it walked funny ... I can't describe it ... it looked awkward ... almost looked like it was floating, to tell you the truth. Very smooth – just glided across the road like it was nothing."

Robert was a very credible witness, and it is interesting to compare his report to that Glenn Kilmartin, who had an extremely close encounter with a similar (or the same) creature at virtually the same spot, on the same road, on 13 November 2018. (Case 269)

In fact, yet another huge ape-man had run in front of a vehicle three kilometres from the Land Warfare Centre on 21 March 1986. The driver, Lester Davison, told us that "The infantry fellows in 10 RIC have seen quite a few [of the creatures] in the Land Warfare Centre." (*The Yowie*, p. 84).

Gilston Road, where Aaron Carmichael collided with a yowie in March 2001, is less than five kilometres east of the Land Warfare Centre. (*The Yowie*, pp. 67-68).

Robert's previous experiences

Robert thinks he may have been close to yowies on a couple of occasions prior to his 2015 encounter. On his property at Bottle Tree Lane, Beechmont, he has twice heard unexplained heavy footsteps and has had big rocks moved, apparently thrown, around. One rock that was shifted was immensely heavy – one metre in diameter!

He mentioned, also, that residents of Numinbah Valley, about two kilometres south-east of Beechmont, told him that a yowie they called "Big Red" had often been seen scavenging through the local tip.

Case 253. 2 Nov 2015. About 5 km north-east of Meckering, WA

Jeannie and Jim contacted AYR on 31 Dec 2015 and were interviewed by Tony Healy on 3 August 2018.

The incident occurred as they were driving from Perth to their property at Gabbin.

> **Jim:** "We'd been driving down the Meckering to Dowerin road for 10 to 15 minutes, [not going fast] because we were travelling at night.
>
> "It was on a long, straight stretch of road. We had the high beams on, I've got those big LED spotties – they light up everything just like daylight, so it was crystal clear.
>
> "We saw this thing standing on the [left-hand] side, staring off into the paddock ... it was *way* bigger than a man, it was just ... when you consider that our vehicle, an Isuzu D Max 4WD, is six-foot tall ... this thing was way bigger than that.
>
> "The arms, legs, torso were just *huge*, and it was light in colour, grey."
>
> **Jeannie:** "The hairs on our arms raised when we saw it – we thought, 'Oh no!' [laughs] It was scary! It was looking into the paddock, so it had its back to us – but we saw the shape of it."
>
> **Jim:** "It was probably only three or four feet from the car as we went past. I was only doing about 50 to 60 kilometres an hour – and we got a good look at this thing.
>
> "We looked at each other and said, 'What the hell was that?' And my wife was saying, 'Let's go back!' And I'm, like, '*No!*' [laughs] I don't tempt fate. You

Jeannie and Jim

know, this creature was *big*, it really was. It could have been seven or eight foot … really, *really* big. I don't scary easy, but something like that … I'm not going to turn around.

"You know when people talk about a big man with shoulders the width of an axe handle? Well, this was way bigger than that – it was huge. The arms were really long, and the legs. Its torso was really long. The hair on the body was really long. But my wife probably got a better look at it, because it was on her side, and she was almost right next to it.

Jeannie: "It was hairy all over and creamy in colour. Its neck was all covered in hair as well – you know those pictures online of the yeti? It looked a bit like that. And its arms were hanging down and they were really long. It was *huge.*

"It had wide shoulders and a sort of smaller head, but it was all hair. The palm of [one] hand was facing back towards me, and it had, like, black paws – black hands.

"It didn't turn around. It was looking into the paddock at the sheep –as if it was going to go and pounce on some of them. Afterwards, we phoned up the Shire to tell them what we saw, thinking that it might attack the sheep. But they hadn't heard of it before."

The area they were driving through is "all open, undulating country … wheat belt country; we don't have any mountains here. There's not really any forest around."

More black panthers

Interestingly, Jeannie, like a significant number of other yowie eyewitnesses, has seen another of Australia's elusive cryptids:

"We do a lot of country driving. I've even seen a panther out here – have you heard of that?"

Tony: "Oh yes– I'm very interested in big cat sightings – the 'Cordering Cougar' and so on."

Jeannie: "Yes, that's it – I've seen it! It was at 10 o'clock one morning, about a year ago. Oh, I know when it was: 5th of January – I was going down to Perth to have lunch with my daughter for my birthday.

"I was on my own, driving slowly, and all of a sudden, this big black cat – well, a panther – he had that long curly tail – went from one side of the road to the other in just one jump!

"That was on East Boundary Road – just down the road from our farm at Gabbin.

"My neighbour has seen it, and his son saw it drinking from their dam. Another friend who has a property down the road was out shooting one night with his son and grandchildren and he got it right in front of his gun to shoot it, but he froze – and this was just recently."

Case 254. 2015, 2016 and 2017. Near Broke, NSW

Stephen Kelly contacted AYR on 10 May 2017. Paul Cropper interviewed him and his daughter Kayla on 12 May and visited the site with Stephen nine days later.

Over the course of 28 months, Stephen encountered what appeared to be the same yowie on three occasions, on the same stretch of road. The first incident occurred two days before Christmas 2015 as he and 21-year-old Kayla were driving home from Maitland to Broke.

First sighting, 23 Dec 2015

> "It was between 9.30 and 9:45 pm, and we were coming back on the Broke Road. I do a lot of mine work, and drive that road twice a day, every day.
>
> "There was a fairly large thunderstorm happening, so the place was lit up fairly well, and we were coming over the hill. On the right-hand side [at the bottom of the hill] there's a big dam, where you often see kangaroos, pigs, and wild dogs. As I came down the hill, there was this thing standing slap bang, right there on the left-hand side, on the grass right next to the [bitumen].
>
> "I thought, 'Oh, a kangaroo,' but it didn't move, and then I thought, 'That's no kangaroo – that's pretty big!' And my daughter said, 'What the hell's that?' And I said, 'I dunno!'
>
> "Then it stepped back about a metre towards the tree line. When we went past, it [was] only about three metres from the car. It was very tall – we were driving a Ford Courier 4WD, which is pretty high, and its head and shoulders would have been above the roof line.
>
> "My son Kyle is six foot six, and he would have come up to about its shoulder – so it was probably seven and a half foot, something like that.
>
> "As I passed, I looked straight at it, and then in the rear vision mirror. The [vehicle] had work lights on the back – a light bar – and I could see [that the creature] didn't move away but had sort of turned to face the vehicle. And that was the last we seen of it. All up, we probably had it in view for about 25 to 30 seconds.
>
> "I got a fairly good look at its upper torso. If it was a human, you'd say he'd been on the steroids – because he had a very muscular chest. Fairly wide across the shoulders. He'd be one and a half times my chest size, easy. So he'd be two and a half to three foot across the chest – really wide.
>
> "It was a human shape, but not V-shaped like a human who's been doing weights. This was wide right down to its hips. It was definitely not a human, way too big for any human I've ever seen.
>
> "The main thing I remember thinking was that there wasn't much of a neck.

Big, muscly shoulders with no neck – the head just sort of joined straight onto the shoulders. The eyes were fairly large: about as round as [the bottom of] a Coke can. A fairly elongated head – a wide, long, sort of head. A big forehead – from [its eyes] to the top of its head was about twice that of a human. I didn't make out much of a nose or mouth.

"It had dark ginger hair; not a lot on the chest – there was a bit of light reflection on the chest. But on the shoulders and head it was fairly long – shaggy, like a mangy dog, probably about 10 to 15 centimetres long."

Stephen Kelly at the Broke Road site

Second sighting, June or July 2016

About six months after the first event, Stephen saw what he believes was the same creature – in almost exactly the same spot.

"It was winter, just on dusk. Within 100 metres of where I seen it the first time.

"I came down the hill again – it's a long hill – and two kangaroos bolted straight across the road, which is pretty common, so I took my foot off the accelerator, and [the yowie] was on the right-hand side of the road this time.

"There's an old dam, with an old pump station there, and [the creature] was probably 25 to 30 metres from the side of the road, right next to the dam. I didn't see it front-on this time, because it was walking towards the dam.

"It was around about 5.30 pm – not quite dark, but not real lit-up either, but because of the clearing next to the dam, [the yowie] stood out like dogs' balls, as they say – I could see it as plain as day.

"It wasn't a very long [look] that time, probably just 10 seconds. I was looking at it back-on. It sort of stooped a little bit, didn't walk bolt upright. And its arms came down towards its knees – really long arms.

"I've often thought that I should have stopped that time and went in and had a look, but I just slowed down a bit because it's a bit of a busy road that time of day – with the opening and closing of the pits and shift changes – and there's not really anywhere to pull off the side."

At the time of our interview, he hadn't ventured in to look for tracks because the dam is on Army land – "They'd put a bullet in you." [Laughs]

Third sighting (probable), 16 May 2017

Stephen is fairly sure that he came close to the same yowie (or a similar one) on a third occasion, about 100 metres from the site of the earlier incidents.

"It was late at night. My wife was with me; we were coming the other way – going *up* the hill. I had the high beams on, and we saw a fairly tall shadow on the road ahead. We didn't get a spot-on look at it, but because the shadow was so long … I know you get a bit of parallax error – but it was a very long shadow. We're used to seeing 'roos through there – it was definitely not a 'roo."

Stephen went on to mention something else that may have been connected to the yowie:

"That night, too, there were a hell of a lot of rocks on the road. They were about twice the size of a man's fist. There's no hill where they could have rolled from – that seemed a bit weird, because there were quite a few of them."

On 21 May 2017, when Paul was guided to the site by Stephen, he noticed that, within 100 metres of the stretch where the yowie was repeatedly seen, the road is crossed by a double row of high-tension power lines, supported on huge pylons.

Quite a few other yowie sightings have occurred very close to power lines, microwave towers and other electrical infrastructure (e.g., Cases 214 and 232). The fact that Stephen's first, and best, sighting occurred during "a fairly large thunderstorm" might also be significant.

Case 255. April 2016 and Feb 2018. Bell Thorpe and Conondale, QLD

"J" (name on file) contacted AYR on 7 Aug 2018 and was interviewed by Tony Healy the following day.

First sighting: April 2016. Near Bell Thorpe, about 15 km north-east of Kilcoy

"J": "I was with a friend who has a property there. It was about 4:30 to 5 in the afternoon; the sun was getting a bit lower. We were up there, just doing a bit of photography [making a record of] the size of the deer and so on. We do a bit of hunting on his property, and we were trying to see how many deer cross from there into the national park [Conondale National Park].

"We didn't see anything, so we put the camera away and were coming back along 'Postman's Track' which comes to his place, and we saw, about 40 to 50 metres ahead, something crouched beside a tree – we could only see its back.

"We didn't walk directly up to it, we sort of walked around it – about a 40 to 50 metre radius, and when we got closer it stood up. It was probably eight-and-a-half to nine-foot tall, really heavy build … twice my shoulder width across – about three foot across. The arms on it were long.

"The first thing we thought was that someone had gone out to sort of put the wind up us, but it moved too freely for it to have been someone in a costume. It was sort of hunched over and had very long blackish-brown fur. And it disappeared as quickly as we saw it.

"I went down to have a bit of a look, but because of the type of ground there were no actual footprints, although you could see where the grass and the shrubs had been pushed down.

"So, the next day we came back out and couldn't find anything initially, but then you could see some small trees about an inch and a half in diameter had been pushed over – it definitely gave the area a whole new vibe." [Laughs]

Tony: "Was there any noticeable odour?

"J": "Not that we noticed, really. We had [sprayed ourselves] with scent suppressant so we could get close to the deer. And the brand we were using had a bit of a scent to it, so we were mainly just smelling that."

Second sighting: Feb 2018. Near Conondale, about 10 km north of Bell Thorpe

"J": "The second encounter was a bit more, 'How's your mother?' [Laughs]

"Same sort of area, but in Conondale, 15 minutes away [from the first site]. I was at another friend's property; he has just sold it because this experience scared him so much.

"It's very wild country; rough terrain. We were coming back from hunting feral dogs, at about 5.30 to 6 o'clock, we dropped into a bit of a creek ravine and saw something up on the hill – it just looked like a heap of lantana that's gone over a branch. When we walked over to it, it sort of turned and faced us, it was probably only 20 metres away, and we've heard where people have said yowies [can be aggressive] so we started backing away.

"You could see it wasn't so much aggressive – it looked more concerned – anxious, if you know what I mean.

"We watched it for about a minute and it began to back away from us and I handed my rifle to my friend, and he sort of backed up behind a tree out of sight and I just stood there, and it actually started coming towards us, but not in an aggressive manner, more in a curious sort of manner.

"So it came to within, probably, five metres – it was at the point where we could feel its breath.

"The sun was just on its way down, and we were in thick scrub so it seemed like it was a lot later – with limited sunlight coming through [but] I could just about have told you the colour of its eyes.

"It was pretty much the same as the other one except maybe a bit smaller – probably seven-and-a-half to eight feet – and he was dark, almost black. It was as if you got a big silverback gorilla and scaled him up to, like, eight-foot tall and gave him long shaggy hair, that's what we were looking at. Same sort of features.

"From what we could see [the skin] looked black but it could have been a dark brown. The hair on its body was 4 to 5 inches long, but a bit shorter on the head.

"The nose was like a human's, but very pushed in [the nostrils pointing forward rather than down]. The ears seemed too small for its head – almost as if you took an adult's head and put a toddler's ears on it.

"The eyes were mainly black with a little ring of brown in them. [Where a human would have the whites of his eyes] this thing just had black. It had hair down to about an inch and a half above the eyes and there wasn't much hair [on the face]."

He didn't notice its mouth, teeth, or jaw.

"Its hands were four or five times the size of mine. There was hair on them, and the hands and feet looked virtually like those of a large gorilla – very apelike."

Tony: "Where the toes more like fingers or more like toes?"

"J": "No – they were definitely more like toes, [they] were longer, but I wouldn't say they'd be able to grip anything."

Tony: "It seems you had the impression the creature was male, but did you notice any genitalia?"

"J": "No – I didn't see anything!"

Tony: "So you and your mate could actually smell the breath of this thing?"

"J": "Yes – the breath was foul – I guess if you got a dog that's been eating nothing but dead meat for a while – it had that sort of very strong, potent sort of breath."

Tony: "It's interesting that you and your mate weren't terrified out of your wits – you stood your ground. This thing walked up to you. What did it do once it got there?"

"J": "It was just looking at us. As it walked up, I didn't know whether to back away, but I thought, 'Bugger it, if something's going to happen, there is no avoiding it,' so we just sort of stood there. I was right in front of it for probably two or three minutes. From what I've heard about how fast they are, if it had wanted us there was no way we were going to get away. I was looking at its face and eyes to see [if there was any change in its attitude, any sign of danger].

"As it came over it looked back at the bush where it had been in the first place – like a wild mulberry bush – it looked at that and then back at me, and it saw my mate and then backed away and disappeared into the thickets."

Tony: "I suppose if things had got nasty your mate could have shot the creature."

"J": "He had the rifle lined up through a fork of the tree just in case something did happen. It was his rifle; I think it was a .270. I made it clear to him not to take a shot unless it was the only remaining option."

Tony: "Do you think it was trying to communicate with you?"

"J": "I think so; my mate thinks I'm an idiot. We only spoke about it that one night and we haven't spoken about it since."

Bell Thorpe and Conondale are just 20 kilometres north-east of Kilcoy, a locality notorious for yowie activity, and this report conforms in most ways with the testimony of many other witnesses.

One detail of the second yowie's behaviour was, however, very unusual, if not unique: the way it walked to within five metres of the witness and stood face-to-face with him for about two minutes.

It seems very strange that it would have done so – doubly strange, as it must surely have been aware of the rifle. Equally odd was the witness's reaction – or lack of reaction – at being so close to the huge creature. Unlike so many other "close encounter" witnesses, who experienced abject terror at the time and what can only be described as post-traumatic stress disorder afterwards, he experienced no great fear when face-to-face with the giant, and zero after-effects.

While his report is very interesting as it stands, it would have been greatly strengthened if his friends had been willing to provide some corroboration. Unfortunately, neither of them wanted to discuss it at all.

Case 256. 11 May 2016. Ewingsdale, West Byron Bay, NSW

The following report was shared with us by the well-known naturalist and broadcaster, Gary Opit:

> "My close friend of 30 years, Henning Beth, at 7.10 pm saw a yowie brightly illuminated by his car headlights, 30 to 40 metres away, crossing Bay Vista Lane, left to right, going north through a treed area only 15 metres from houses in an old estate, where each house is built on a hectare of land with lawns, gardens, fruit trees and regrowth forest.
>
> "About 2.5 metres tall, covered in brown, shaggy hair from the top of its head down to its legs, it was walking upright, unhurried and calm. Henning had a clear view of it and observed that its head hair dropped straight down at the back of its head. He noticed that the arms were not swinging but hanging down beside the body as it walked."

The Byron Bay area is rich in Indigenous Hairy Man lore. Non-Aboriginal people, too, have encountered the creatures there from at least as early as 1904. (Case 114 and *The Yowie*, pp. 16 and 215.)

Case 257. July, 2016. Boompa, south-east QLD

A young lady, still on her provisional driver's licence, experienced two sightings just a week apart, at the same time, and at virtually the same spot, on the Brooweena – Woolooga Road. She sent a written report to AYR and was interviewed by Dean Harrison on 7 August 2019.

First Sighting

"I left work at Boompa at 8pm. Drove [south towards Woolooga] for about 20 minutes. It's all open cattle properties [until] the dirt part of the road, [then] it's bush on both sides.

"I had to watch out for cows until I got to the dirt patch. It was there that I saw a very large and muscular, orangey brown creature run across, about 20 to 30 metres in front of me. It scared the living hell out of me. I had a dog with me who also wasn't very happy about what we saw. I leant over and locked the passenger side and drivers side doors – I was in a ute.

"It came from the right side of the road at about 20 or 30 metres in front of me. It was just over 6ft tall. It had a very broad head and not much of a neck that went onto very large solid shoulders. It had very large hands and arms. The hair was dirty and wiry looking. I could clearly see where its back went down to its buttocks and went onto very muscular legs, like a body builder's legs. The hair was a browny orangey colour, but darker than an orangutan and it was 2 to 3" long. The hair length varied in certain places. The arm's reached down to its knees. Where we have a rounded head, it had what looked like a crest that came up the back, like an ape's head.

"I didn't see any eye shine because it was side-on when it ran across the road. It went up the hill into the bush.

"Because of how the excavator had pushed all the dirt up on the sides of the road, I think it was already coming down [one side] and had to keep going and get up the other side or I would hit it."

Second Sighting

"Then, a week later [again leaving work at 8pm and driving along the same road] not more than 50 metres away from where I saw the first one, there was another one ...

"I had my doors locked due to the last time. [The second creature] looked very similar to the first one, but about a foot and a half taller and with more muscle. It was like it was waiting for me. It had come down off the embankment and was standing on the right-hand side, like it was waiting for me ... standing with its arms down to its side. It almost looked like a bouncer on a door ... looked like it was waiting for something to happen.

"I couldn't take my eyes off it, even while driving on a dangerous road, it had me so shocked with fear that I couldn't look away. As I got close, it let off that horrible scream/yell. It sent shivers down my spine and pretty much sent me into tears. I was that scared. My dog [saw it too, and] jumped down off the seat and was trying to get under the seat.

"I tried to speed up to get past it and it chased me for 2 or 3 very large running steps and let off another menacing yell. It was horrifying.

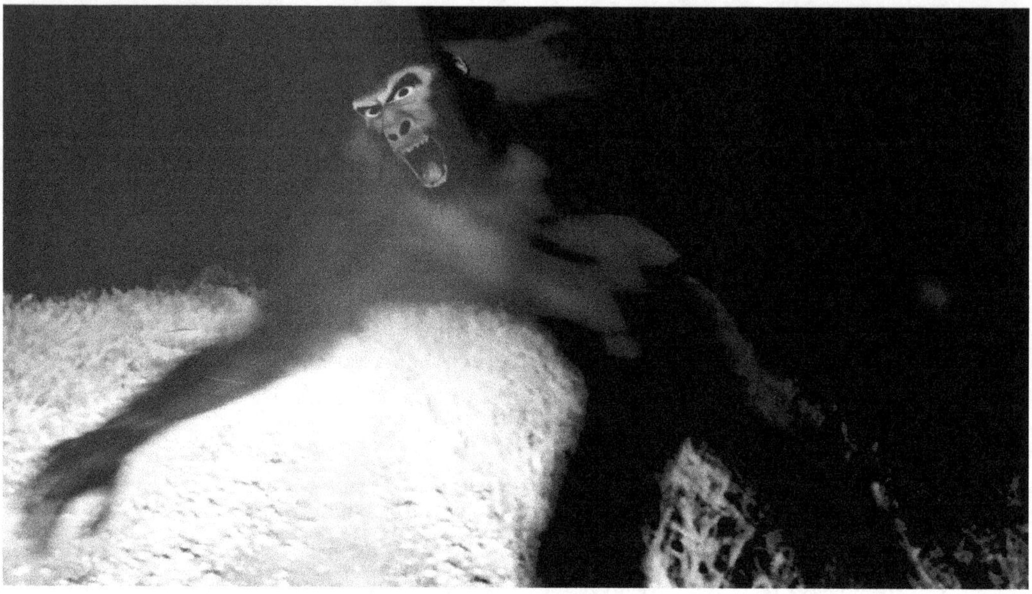

"It had an angry face ... horrifying".

"It was 7ft tall and bigger than the first one in muscle and size. It was the same colour. Hair was the same length. The face was very dark and leathery looking and had a large eye ridge. It had green yellowy eyes, which to me looked very menacing. I could see light coming from the eyes before the headlights hit it. They self-illuminated.

"The nose was wide and flat. The mouth came out like an ape or monkey, like a muzzle. It didn't have much facial hair. The hair was longer underneath its chin than anywhere else. Comparing it to a human age, I would say it looked like a 30 or 40-year-old man. The first one would have been 10 years younger.

"Ape/human/Neanderthal – it was like someone got all three of them and pushed them together. It had features from all three.

"It had an angry face, like when you're warning someone off. When I came up right beside it, it would have been only 2 or 3ft from the driver's side door. That's when it yelled and threw its arms up in the air. It was a low guttural noise. It was like a mock charge and it threw its arms up about parallel with its shoulders and reaching out towards the car. The hands were like leathery ape hands.

"It was horrifying. After I went past, I was shaking, and I went over scenarios in my head of what could have happened. I came home rambling and crying. It was very, very scary. My dog was whimpering as well.

"I have been driving that back road with my mother ever since I was a kid. Up until then, I always took that road and felt comfortable on it. I quit work a month later because I was too scared to drive on it. Now I avoid it at all times."

Woolooga is about 15 kilometres east of Kilkiven – site of many yowie encounters.

Case 258. 15 July 2016. Harrietville, VIC

Jarrod Mitchell contacted AYR on 18 July 2016 and was interviewed by Paul Cropper a week later.

The Mitchell family has lived on a property near Harrietville for many years. One afternoon, as 24-year-old Jarrod, his father and uncle, were deer hunting on a pine and scrub-covered hill, they encountered an enormous, hair-covered ape-man.

Jarrod said that between 4.30 and 5.30, with about an hour of light left, they were walking over a little embankment when they heard "guttural" sounds, a very low frequency growl and heavy, running footsteps. Then they saw, just 40 to 60 metres away, standing next to a tree, a "large bigfoot … about eight-foot tall – it was *huge*, mate!"

It was covered in dark hair, "either real dark brown or black," with massive shoulders and arms that were extremely long – "the proportion was all wrong – [they extended] down to about its knees." It had no discernible neck.

"Its eyes … I knew it was looking at us … its face … was more like a human's than a chimpanzee's ... not like 'Harry and the Hendersons.' Not a cone head – just a rounded head, but the body was not human."

The face-to-face encounter gave Jarrod "a weird, strange feeling almost as if I was paralysed." Although they were well armed – Jarrod had a .22 and the older men were carrying .30-30s – they had no thought of shooting the creature, which stood looking at them for three to four minutes before turning and simply walking away.

The episode shook Jarrod quite profoundly, but he noticed that it affected the older men even more – his father's face had turned quite pale, and "my uncle was shaking, just shaking."

Soon afterwards, as Jarrod began to discuss what they'd seen, his dad said, "Shush – just shut up – don't talk to your mother about this!"

Since then, "Dad has been very quiet [as if he's] in denial. He hasn't been to work; mum is worried about him because he's gone all quiet."

Jarrod's uncle, however, has been a little more forthcoming, revealing that he'd heard what he now realised were the yowie's guttural vocalisations once before. On that occasion, his dog barked and ran away. He found it later – dead.

At least one other yowie event has been reported in the Harrietville area – in 1987 three campers were supposedly terrified by a huge creature that rocked their caravan three nights in a row. We never got around to verifying that particular story.

Case 259. Oct 2016. Near Goongerah, VIC, and Oct or Nov 2016, near Chinchilla, QLD

After experiencing yowie encounters in two widely separated locations within the space of just three weeks, 29-year-old Sam Keown contacted AYR on 8 November 2016. He was interviewed by Paul Cropper on 10 November 2016 and by Tony Healy on 14 Jan 2018.

All yowie reports are, almost by definition, strange, but some are so odd that they can only be categorised as "high strangeness" cases. Sam's experiences fall into that category.

First sighting. Early October 2016. Near Goongerah, VIC

The first incident occurred about 30 kilometres south of the New South Wales/Victorian border, on a stretch of Bonang Road that cuts between Snowy River National Park and Errinundra National Park – an extremely wild, mountainous area. The nearest settlement, Goongerah, is about ten kilometres to the south.

Sam: "I'm a timber cutter and machinery operator, and I work between Victoria, New South Wales, and Queensland.

"It was about 10 pm, I was driving south towards Orbost; it's a dirt road, pretty windy. I've got a Toyota Hi Lux ute with a bar light on the front, and I had my high beams on, and I saw something up in the bush, on the embankment to my right. Then it ran across in front of me, and looked at me, and kept running to the left-hand side [and down a bank]. I only missed it by probably a metre.

Sam Keown

"It would have been about 10 feet tall – I had to crouch my head down and look up to see its head. Really, really broad shoulders and chest. I'm nearly six foot four and weigh about 110 kilos; I'd say this was 200 kilos, maybe even more.

"It was hairy, a light, reddish-brown colour. Long, shaggy-looking hair. Really long arms, and from what I could see, the hands were huge, *massive*.

"I had the feeling that it looked startled. It looked kind of like a humanoid. The face was almost like a really deformed, out-of-shape, human sort of face, and the head from the forehead up, went into a kind of conical shape. The skin on its face looked tanned – not really dark – but like someone who's worked in the sun for 30 years ... really big eyes and – it must have been the reflection from the lights – they were a sort of yellowy-orange. I didn't see any teeth.

"I had the window down, and there was a horrible, horrible smell, almost like a rotten egg gas, like sulphur. The air was just rich with the smell ... it was just *unbelievable*.

"I pulled up in the middle of the road, about 50 metres further on. I couldn't believe what I'd just seen. I sat there trying to find where exactly I was [on his phone's GPS].

"And then – I've got bright LED lights on the back of my ute – in the rear-view mirror, I saw this thing running back up, out of the bush, onto the road. It was groaning, grunting, real loud. It looked angry and was running full pelt towards me!

"So, I put the car into gear and my foot to the floor – and this thing kept up with me, I could hear it grunting and snorting. When I reached about 80 kilometres an hour, I finally lost it. [And at that point] it let out this almighty sort of *screech* – almost like the screech of a cockatoo but crossed with a roar – and *extremely* loud. I can't think of anything that could come close to it. And I thought, instantaneously: 'Death!'

"And ... Shit, mate ... I'm getting goose bumps just talking about it!

"The whole thing lasted, probably, no more than a minute, minute and a half."

Second sighting. October or early November 2016. Near Chinchilla, QLD

Sam: "This was only about three weeks after my first encounter.

"It was about 7:30 at night; I was on a dirt road near the Barakula State Forest, roughly between Miles and Chinchilla, driving quite fast – 70 to 80 kilometres per hour.

"I had my high beams on, and something walked out, on all fours, from the left, about 50 metres ahead. I thought at first it might have been a pig. I beeped the horn [and it looked] quite startled – and stood up and looked at me.

"Everything happened very fast. It was on the verge [but then] started walking into the middle of the road, moving like a person. I couldn't avoid it and clipped it with the left-hand side of my bull bar [and] it fell off, back into the bush [making] the same sort of screeching, roaring noise as the other one had – quite loud. I felt it through the closed passenger-side window: I could feel the vibration of it through the car!

"I went to jump on the brakes, but then jumped back on the accelerator, went down a gear and started accelerating again – I just didn't want to be there! [laughs] And I didn't stop to look back to see what had happened to it – I was quite scared. I wasn't stopping at all, mate!"

The creature was very similar to the one he'd seen in Victoria, but perhaps a bit smaller: "It would have stood, probably, between seven and nine foot … very hairy … really wide shoulders. Massive, massive arms. It would have weighed an easy 200 to 250 kilos.

"Its body was covered in reddish-brown hair, apart from maybe part of its feet. I didn't notice how many toes, but I did see a big toe, that's all I remember. The tops of its hands were quite hairy; it looked like it had five fingers and a thumb.

"It had characteristics of a man, but its face had very apelike features, [it] was quite long … I don't know what to compare it to, but it opened its mouth and I know it had a really big set of teeth. I didn't see how many, but [they were] like fangs. All I saw was, like, a mouthful of fangs – like a piranha. [Laughs]

"There was a lot of hair around the cheekbones and the eyes, but part of its forehead was exposed. [It was] a very dark tan.

"Its nose was similar to, say, a gorilla's: pushed back, so its nostrils weren't pointing down: they were pushed forward."

Weird eyes

"The eyes reflected a reddish colour – you know, like the colour of a red moon? Very similar to cats' or foxes' eyes. And they blinked sideways – its eyes blinked *sideways* at me.

"I'm very positive that [the eyelids closed from the side]. I found it a bit surreal because [the creature I saw] during the first encounter also blinked sideways at me, and I immediately thought, 'No, this is just not normal!' Only [the eyelids] of reptiles … crocodiles [open and close] sideways.

"I thought, 'Oh no! This can't be happening again – not again!'

"I didn't sleep very well for a good two to three months. It was horrific."

In view of what he said about the creatures' extremely strange eyes and given the apparently astronomical odds against anyone experiencing two such similar encounters in locations

1,200 kilometres apart, we asked Sam if he thought there might be something "uncanny" about the yowie phenomenon.

> **Sam:** "I thought … it's something that can't be right, seeing the same type of creature twice within three weeks to a month, I actually started questioning myself and my actions, sort of … I started thinking.
>
> "I'm very cautious now … I don't go out at night by myself anymore. When I'm hunting, I always go with two or three other people."

Apparent psychic ability

Asked if he'd had any other strange experiences, he replied in the affirmative: "Yeah, I have actually, mate. From a very young age I've always … I dunno, I've always seen … I've seen ghosts …. I can always feel when there's something not right."

Other strange animals

Like a significant number of other yowie witnesses, Sam has seen another type of Australian cryptid:

> "Yes, I've seen what appeared to be a Tasmanian tiger [thylacine] out the back of Singleton, that was probably 10 or so years ago [c. 2008].
>
> "All I remember was that it had very faint black stripes on its back, it was bigger than a dingo and its head very much resembled [that of a] Tasmanian Tiger. I saw it from about 20 metres away, running through the bush. I didn't bother pursuing it."

When another notorious Australian cryptid, the mysterious, so-far-uncatchable "black panther" was mentioned, Sam said that although he'd never seen one, his father had – in the Great Alpine National Park just near Licola, Victoria, in the 1980s.

Aboriginal lore

> "My mother was a teacher at Woolgoolga on the New South Wales north coast, and I grew up there. I'm a white Australian – but I kind of grew up, in part, in that Aboriginal culture. And part of the culture is that they are very much in touch with the land, and with everything that goes on around them.
>
> "The blackfellows always talked about a spirit called the junjudee man. They said that it basically looked like, sort of, an old man in a six-year-old body."

Junjudees

> "I was working on a logging coup down in south-east Gippsland, and the bottom of the coup looked out on a rocky area – the Buchan Caves.
>
> "We were having lunch near the caves, and there was something like a hole in the top of a rock, 50 to 70 metres away, and this thing popped its head up. At first, I couldn't make out what it was.
>
> "It jumped up onto the top of a rock and another three jumped up. They were little man-shaped creatures, like people, but with hair [all over] their bodies. We couldn't make out whether they were talking [to each other] or not. They looked at us and then disappeared over the rock, heading away.
>
> "And the bloke sitting next to me dropped his coffee cup!" [Laughs]

He has heard from Indigenous people "that the junjudee man is quite an evil spirit. So is the yowie – they say those are things you don't go near or tamper with.

> "I'm a contractor – a machinery operator, in forestry – I'm qualified to do specialty work in highly environmentally sensitive areas, so I've spent a lot of time out in the bush – I go everywhere and anywhere – I was doing some work up in a range near Toowoomba and I was talking to some forestry heritage people, Aboriginal people, up there. This was after my first yowie encounter – and I was telling them about the yowie encounter, and they just went *white.* They said, 'It's very rare that we meet a white feller who's seen one. We fully believe in them – they're very real – but for a white feller to see one, that's pretty amazing.'
>
> "After my second encounter, I went back and got in touch with them again and they know all that land out the back of the Darling Downs – the Surat Basin – and unfortunately a lot of the land out there got destroyed because of gas exploration – a gas pipeline going to Gladstone, etc.
>
> "They basically said, 'Seeing [a Hairy Man] is, like, an awakening. If we know that there is one nearby, we don't wait around – we just leave the area'. So, in their culture a yowie sighting is considered a signal that something is wrong – you know, like in the environment – when frogs start dying in an area?"

Sceptics might dismiss Sam's report as the mother of all tall stories or as the product of a fantasy-prone mind, but we don't doubt his honesty or sanity.

As he himself has acknowledged, however, "seeing the same type of creature twice within three weeks [is] something that can't be right … I thought, 'Oh no! This can't be happening again – not again!'"

He also emphasised how "surreal" it was that the eyes of both creatures "… blinked *sideways* at me … I immediately thought, 'No, this is just not normal!' Only [the eyelids] of reptiles … [open and close] sideways."

Sceptics can chortle as much as they like, but in Sam's defence we should remind readers that unfeasible, sideways-opening eyes have been described once before – by Petra Van De Moosdyk, who had close encounters with yowies at Keiraville, NSW in the late 1990s. (Case 223)

Given the many strange and interesting elements in his testimony, it seems highly likely that Sam's yowie encounters were psychic experiences of some kind.

Case 260. June 2017. Border Ranges, northern NSW

This case is particularly interesting because it involves a highly qualified scientist who is very unlikely to have misidentified what he saw. Because of his qualifications, because of his familiarity with the area in question and because he was so very close to the creature, we consider him to be one of our "star witnesses."

He contacted AYR on 15 June 2019 and was interviewed by Paul Cropper shortly thereafter.

His remarkable encounter occurred on a winter's day at around 3 pm, while he was involved in field work on the New South Wales side of the Border Ranges.

> **Witness:** "I ventured up a steep escarpment creek … as steep as you could ever imagine, for a bird's eye view, and up a basalt cascade to a smaller cascade. I stopped there to work out how to navigate it, and to the right, within ten metres of me, was a huge, crouching creature.
>
> "It was covered in long, shaggy, reddish hair; almost auburn. It had a grey, ape-humanoid face and huge eyes – as big as golf balls; like fish eyes – staring straight ahead.
>
> "It sounds bizarre, but I got the distinct impression it was trying to conceal its presence by pretending to be a tree or some inanimate object – like a child pretending you can't see them, while crouching right in front of you. It obviously saw me before I saw it, and decided to squat there and stare, like a marble statue."

It was sitting with its knees drawn up in front of its torso, with its arms wrapped around its shins.

> "Because it was crouched, it was difficult to say exactly how tall it was, but … from how the knees were poking up from the crouching position, with the feet in front of him, and the bulk of its frame, I could tell it was bigger than me. I'm around the six-foot mark. It was tall.

"I've read some of those [yowie] reports where they are muscular and huge, something along the lines of a gorilla, but this wasn't anything like that. I wouldn't [call] it lean, it was more along the lines of a bigger-built orangutan, probably because of the colour of its hair. It had no apparent neck – its head was coming out of its chest. It had hair on its head, but the face was totally hairless.

"The face was not that of an orangutan, it was more human than that ... almost humanoid but with some ape-like features, like the nose and the way the mouth came out – the mouth had a muzzle. The area below the nose and around the mouth was puffed up – not necessarily protruding, but certainly not lips like a human being.

"It had ape qualities, but human qualities too. Even something about the giant, fish-eye, golf ball-sized eyes said something human to me as well. It's hard to describe because I'd never seen anything like it before.

"It was slightly higher than me, sitting on the other side of the waterfall. I looked at it, looked away, got spooked, and looked back at it kind of sideways. I had a clear view of it.

"It spooked the hell out of me, to be honest. I didn't get the impression there was any immediate danger, but the experience rattled me pretty heavily. I pretended I didn't see it, turned my back, and went down the mountain, not running because I didn't know what to expect – and just kept going. Once I got away from that part of the mountain, I picked up my speed.

"I had a camera in my kit at the time, and you would think that someone in my position would get the camera and start taking shots, but I didn't, I suppose because of the shock and fear. I was alone and was taken by surprise – I was just so spooked that all I wanted to do was turn around and walk away. The camera was literally the last thing on my mind. Now, unfortunately, the time has passed.

"I have told some family members, but I certainly wouldn't tell any colleagues about it.

"The sighting still unsettles me; not [only because of] the huge, crouched creature, but because its very presence is 'impossible' according to all current and past scientific theories. It disturbs me on many levels.

"It sat with me for the last couple of years, so much so that I [did further study in] palaeoanthropology. I wanted to understand what I had seen and what [could be] living here on this continent.

"I would say unequivocally that it was not a hominin, but rather a hominid. Hominins are anatomically like modern humans, Neanderthals, Denisovans, *Homo erectus*, *Homo Floresiensis*, etc. A hominid is far closer to the ape family.

"I've seen reconstructions of [various extinct] hominids and some of them, I suppose, come close, but this was something different. Scientifically, it shouldn't be here, but even if it was, it should have been here 200 to 300 hundred thousand years ago – not in 2017!"

A similar case

There are many interesting details in this report, notably the posture of the creature, and its rigid, statue-like immobility. In our earlier book we documented a very similar incident during which a Hairy Man squatted, staring straight ahead, in the same manner, as it was approached by several Aboriginal lads on Fingal Peninsula, NSW in 1991.

Unlike the Border Ranges creature, that one finally reacted, but only when one lad, Kyle Slabb, was within arm's reach: "… he looked up at me … the expression was, really, just *blank* … completely blank… and I was speechless … [Kyle's friend] Ernie just screamed, and … everyone ran!" (*The Yowie*, pp.78-82)

After we sent him a copy of the 1991 report and accompanying illustration, the Border Ranges witness was "… amazed at the drawing – that is definitely an identical posture, and his report of the blank stare, until looking up at him, certainly matched my experience (except for the eye contact)."

Kyle Slabb's sketch of the Fingal creature

Case 261. 2017. Nana Glen, 25 km north of Coffs Harbour, NSW

Darren Azzopardi contacted researcher Gary Opit in early March 2017. After Gary forwarded the details to AYR on 14 March, Paul Cropper interviewed Darren and his partner Julia.

In September 2016, Darren and Julia bought several acres just off the main road through Nana Glen. There's bush nearby and big trees along a creek running through the property.

The night before the sighting, they were woken by their two large, well-trained dogs barking excitedly. Normally, they didn't react to native animals.

On the following evening, just before dark – Darren and his partner Julia were walking the dogs on a long strip of crown land that runs up a slope adjoining their property. Over a nearby fence, a horse was grazing in a paddock covered with long grass.

Darren: "We walked up beside the paddock, turned around and came back, and we were just walking and talking – and I saw this *thing*. It was running through the paddock, and we were trying to work out what it was.

"It was getting closer, running on a 45-degree angle past us, because it was heading for these little trees that are nearby. It was [about] 100 metres away, and all I thought was, 'What's this thing got around its head? It had this square sort of head, and as it got closer, we saw its arms. The grass was quite high, so I only noticed the top [of its body]. And as it was running past, Julia said, 'It's running upright – what is it?' And I'm like, 'Bigfoot?' It looked like Bigfoot! [Laughs]

"It came running down the hill at 100 miles an hour; it was moving its arms – big, thick arms – big, thick shoulders – it was really thick and wide and was a light brown colour. It was hairy – like a long-haired dog that hasn't been brushed. When it got closer, I could see the hair on its arms going over its hands. I didn't see its face, but I could see clumps of hair on it.

"I'm six feet tall. This thing was way bigger than me ... but it was so thick – too thick for a human – *huge*. It looked like it would have been way wider than a metre – maybe a metre and a half – hard to tell. It looked square-shaped ... out of proportion. It wasn't a human. I own a gym, and I'm pretty big, but this thing would have made me look like a Chihuahua! It would have been a few feet taller than me, but it's hard to say as it was on an incline in long grass – I wasn't level with it.

"It felt like we were watching it for a long while, but it was probably 20 seconds or so. It was running from my neighbour's property. He's got a plant nursery up there and might have fruit trees."

Darren suspects that the creature was taken by surprise when he and Julia suddenly turned around and retraced their steps Had they continued on, it would have crossed the paddock and entered the trees unseen.

He insists the strange figure couldn't possibly have been a practical joker in a gorilla suit. "No – he was going too fast ... and the detail in the hair ... and for someone to wear a suit and run that quick without falling over would have been impossible. It was running downhill ... a man in a suit would have lost balance and rolled down the hill. It was out of proportion to a human or any animal.

"It was running like Usain Bolt – way too fast for a human. It was just *going*! And as soon as it hit a couple of trees it disappeared. And I'm thinking, 'Okay – we'll get a better look at it when it gets past the trees.'

Darren's sketch of the creature

"Julia went inside the house. She was a bit freaked out by it … and I stayed outside and just stood looking at this tree that it ran behind. I waited there until it got too dark [and didn't see it again]. I think it must have been [hiding behind the tree] that's why I stood there looking at it – I didn't take my eyes off it – but I didn't see anything. But they are quite big trees that have been planted in a line and [the yowie] just could have found its way back to the bush without being seen … but it sort of disappeared into thin air – as soon as it hit the trees it was gone.

"We went to bed and got woken up during the night by a noise that we thought was our alpacas fighting, but they don't make that noise – a weird, yelling sort of noise – really loud – and right in our backyard, just where the creek is. It was nearly a full moon, but when I got up to look, I couldn't see anyone down there.

"We went back to sleep and an hour later the same thing happened – the 'yelling', the dogs barking, so we didn't sleep much that night. We haven't heard it since.

"I went over to the tree next day and stood behind it, [it was big enough to have concealed the yowie] and found a footprint in some cow crap behind the tree. It was basically human shaped. There was an impression of a toe – a big toe impression. That's what it looked like to me. You can't really see it in the photos.

"My girlfriend is the greatest sceptic. I love watching those documentaries [about bigfoot] and she makes me turn them off. But when it happened, she said, 'It's running upright – what is it?' And when she said that, I thought, 'It's got to be a bigfoot! But when you see these things it's freaky … then you think, well, no one really gets close to them and no one gets hurt by them, so they can't be that bad." [Laughs]

Although she was previously very sceptical about things like sasquatches and yowies, Julia's account of the incident corroborates that of her partner.

Julia: "I'm the biggest sceptic – I have absolutely no time for that sort of thing – previous to this! [Laughs]

"When Daz looked at something [in the paddock] and said, 'Oh, look at bigfoot!' I just said, 'Oh God' and rolled my eyes, but then I looked at the horse in the paddock and saw that it was looking at something – and there was this thing running upright. And I turned to Darren and said, 'Why is it running upright?'

"It was very chunky – a big barrel of a body, big arms. I couldn't distinguish a head sitting on a neck, if you know what I mean. If you see a human at a distance, you can distinguish a head, neck and shoulders. But this thing was just like a shape.

"The arms seemed to be pumping, swinging. I could see a fur or hair covering. It looked shaggy, longer than, say, a horse's coat– stringy, shaggy. I interpreted the colour to be caramel blonde."

Strangely calm critters

"I thought, 'Why aren't the horses and cows freaking out?' If I run full pelt towards any of our animals in the paddock they generally spook. But these seemed pretty calm – they were just grazing – they looked at it, turned back, and kept grazing – it was really strange.

"I'm about five foot seven. I think it would have been at least seven foot [but] it might have been eight or nine feet – but it wasn't the height that got me so much as that it was so *thick* – so *wide*. It was too big and moving way too quickly to be a man in a suit.

"I have no idea what it was. I'm as sceptical as they come – but I know I saw something running upright in the paddock and I have absolutely no explanation for it. I don't like speculating, but it was not a misinterpretation – it was not a cow, it was not a horse, it was not a person!"

It is interesting to compare this yowie event to the one reported by Jason Fry in September 2018 (Case 267). During both episodes, enormous, hair-covered, bipedal creatures passed very close to grazing horses in broad daylight without disturbing them in the least.

It might also be worth noting that the way the huge creature "sort of disappeared into thin air" as it ran behind the tree is very reminiscent of the inexplicable vanishing of the small yowie in Jiggi Valley in 2003. (Case 228)

As sceptical readers pounce, gleefully, upon Darren's mention of having watched many documentaries about the bigfoot/sasquatch mystery, we hasten to remind them that Julia, who'd thought such programs utter rubbish, also clearly saw the huge, hairy biped cross their property.

Case 262. 2017. Owens Creek, 40 km west of Mackay, QLD

The witness, "W B" contacted AYR on 12 December 2019 and was interviewed by Paul Cropper in January 2020.

"W B": "I hunt feral pigs. It's strange up here. In the rainforest there's no animals – no lizards, birds, no anything, just a few pigs.

"This night I was missing a couple of dogs (I'd lost six dogs in six months that year), and [around midnight] I was going down the road, whistling and calling out for them.

"In the corner of my eye I caught a small guava tree about 20 metres away, going from side to side ... then I saw this thing beside the tree, whopping it from side to side, while looking at me. I was riding a quad bike with a hunting spotlight, so it was pretty well lit up. It was on the edge of a slashed paddock; visibility was perfect.

"It wasn't much taller than me but was easily three times as wide [with] huge shoulders. It could have been 200 kg, possibly more. Long, messy hair all over. Its hands and face had shorter hair, while the rest of the body was shaggy and knotted from head to toe, a dirty, dark brown with a red tinge.

"It had gripped the tree about halfway up and was whopping it – the top of the tree was beating the ground from side to side. It was doing it to get my attention, I think. Maybe I was near its family, and it was trying to draw my attention away. It wasn't trying to hide.

"We looked at each other for what seemed like three minutes, then it took off, lightning fast, knocking trees over as it ran. Then I could hear him coming back around behind me, doing a big loop, and that's when I felt a fear like I've never known. I was armed with a 12-guage pump action shotgun full of solid rounds, but [it] was not going to be of any use.

"That's when I bailed – I took off flat out, faster than I'd ever ridden before … too afraid to look behind … things were bouncing everywhere, flying off the bike.

"It took about three hours of questioning from my missus before I told her what had happened, and never told another person for two months. I have lived in the mountains here for 20 years and spent many nights alone in the bush, hunting pigs, which I have done since I was 10 years old. I consider myself a very experienced hunter, but I was pretty shaken up – really fearing for my life."

Sketch by "W B"

Case 263. 6 Sept 2017. Wingello, NSW

Sophia (surname confidential) reported the incident to AYR a couple of hours after it happened and was interviewed by Tony Healy four days later.

Wingello is situated on a two-lane road connecting Bundanoon and Tallong. The road runs parallel with a railway line, and, when it passes through Wingello, forms the main street of the village. Within the village, it is sign-posted as Railway Parade.

Sophia lives with her father on Railway Parade. Across the road (and the parallel railway line) is a vast area of bush – Penrose State Forest. She believes that their small residential block has been visited by a yowie, or yowies, on at least two occasions.

The sighting

At about 1.30 am on Wednesday 6 September, 24-year-old Sophia, at home alone, heard what sounded like heavy footsteps moving up the grassy driveway on the eastern side of the house.

Going out the back door to investigate, she immediately noticed a very foul smell, "like burning" and heard something run back towards the front gate. When she was in a position to look down the driveway, she saw the intruder crouched beside a power pole between the front gate and the road. At that stage it just looked like "a big, black ball."

She thought for a moment that it might have been a large wombat, so, switching on her mobile phone's flashlight, she walked to within about 15 metres of the whatever-it-was.

To her alarm, it then stood up, towering to an immense height, and turned to face her. By the light of her phone, augmented by partial moonlight, she could see it had enormously broad shoulders – about four feet wide, long legs, and arms that hung down to about knee level. Its large head was clearly silhouetted, and its eyes, reflecting the torch light, were orange coloured. Although no genitalia were evident, the creature, given its general shape, seemed to be male.

Sophia demonstrating how the yowie crouched

Sophia retreated around to the back of the house, and walked through to her father's bedroom, which overlooks the front yard and street.

From there, she watched as the hairy giant crossed Railway Parade in just two or three huge paces stopping beneath a gum tree on the opposite verge, between the road and the railway line.

It remained there for long enough for Sophia to see that the top of its head was almost level with a particular branch, which she later established was about nine feet above the ground.

The creature finally turned and disappeared into the darkness, apparently crossing the railway line and entering Penrose State Forest.

A large footprint

On the following morning Sophia found what looked like a large footprint in an area of soft grass in the front yard. Because the grass was wet with dew, the imprint was fairly clear, and she could make out the suggestion of a toe pattern, but it was difficult to photograph clearly.

When Tony saw it, four days after the event, it was very faint but still had the general shape and size of a bigfoot-type track – about 18 inches (46 cm) long and 8 inches (20 cm) wide.

The footprint was right next to a small garden bed in which the soil seemed to have been disturbed by something – possibly the yowie.

Some heavy garden furniture – two or three cast iron chairs, in the front and side yards, and even in the back yard – had been tipped over. They could not have been toppled by wind or small animals.

Sophia's sketch

Earlier incidents

Because the iron chairs had been pushed over on a previous occasion, Sophia believes the yowie entered the property at least once prior to 6 September.

Also, on 2 August, she'd recorded some very loud, screaming vocalisations, the like of which she'd never heard, that seemed to come from Penrose State Forest – a huge area of bush on the other side of the railway line.

Naturalist Gary Opit, who listened to the recording later, found the vocalisations interesting and unusual. He didn't think they were made by a fox but couldn't say for sure what they were.

A history of yowie activity

It is very interesting to note that another (or the same) yowie was seen standing beside Railway Parade, just past the outskirts of town towards Bundanoon on 19 November 2011. (Case 244)

Although she was well aware of the yowie mystery (she had once looked through a friend's copy of our 2006 book *The Yowie*) and knew that such creatures had been reported in New South Wales, Sophia hadn't heard of the 2011 sighting, just one kilometre up the road from her own. She was also unaware of the numerous reports that have come from Moreton National Park, a vast wilderness just south-east of Wingello.

In fact, the area around Wingello has produced many yowie reports since the colonial era. (e.g., Case 49)

Case 264. 16 July 2018. Richmond Range National Park, south of Woodenbong, NSW

David McMaugh contacted AYR on 22 July 2018 and Tony Healy interviewed him the following day.

David and his wife Susan, prosperous Queensland-based vintners. were enjoying what they hoped would be a relaxing weekend in a comfortable cabin in Richmond Range National Park.

> **David:** "Susie and I went down to a really nice little off-the-grid cottage up in the range and stayed there Saturday and Sunday nights. It's very isolated – 20 Ks from the bitumen road, and the only people we saw the whole time were two guys on trail bikes. It's the middle of bloody nowhere – a fabulous place. Magnificent views and all that stuff.
>
> "On Sunday night, there was quite a strange noise – I'm sure you've heard possums and koalas – that very loud, grunting sort of noise – and I'm sure you've heard cattle. I'm a farmer, and when you separate cows from their calves, they'll bellow all night. But this was nothing like that; it was sort of a moan. I had a couple of years in New Zealand, doing a bit of deer stalking, and it definitely wasn't a deer bellowing either.
>
> "It spooked Sue, and we thought, 'Gee that's strange – there's someone there!' And these loud, low moans repeated a couple of times.
>
> "So, I got my very powerful torch – the beam goes out to about 500 metres – and walked right around the cottage and out beyond that. Couldn't see any eyeshine anywhere [but] definitely smelt something – like bad B. O. – like someone hadn't had a bloody shower for a few days, you know? [Laughs] There's a feedlot on a property opposite ours in Boonah, and it smelt a bit like that.
>
> "The cottage is in the middle of a paddock – the trees at least 150 metres away on one side and hundreds of metres away on the other sides. There was no stock there, no possums, no koalas, and I thought, 'That's bloody odd!'

"We left at about 8 am. It's a well graded forestry road, but relatively dangerous. I was driving an XR 6 ute and going pretty slowly – only about 20-25 Ks. We were going down some very steep-sided spurs, so I was seriously watching the road. We'd just come out of primary rainforest into open, eucalypt forest, and – *pow* – there was this fleeting figure running across, about 40 or 50 metres in front of the vehicle.

"I immediately said to Sue, 'Hey, hey – did you see that?' And she said, 'What? What?'

"I said, 'There was someone running across the road!' It was all over in a couple of seconds.

"I'm a reasonably sensible guy. I definitely saw this upright figure run across. What caught my attention was it was quite wide across the shoulders or the body. And I noticed what I thought was an arm going up to brush a branch or something away.

"It was about the height of a child – maybe five foot ... a brindley-tan colour, and [roughly human shaped]. There were definitely legs. It moved pretty bloody fast too – in a matter of seconds it covered 20 metres or so.

"We pulled up and looked around, but didn't see or hear anything, and I said to my wife, 'I reckon that was a yowie!'"

Case 265. 25 July 2018, Carnarvon Highway, north of Roma, QLD

Mandy Psimaris contacted AYR on 27 July 2018 and was interviewed by Tony Healy the following day.

Mandy: "It occurred last Wednesday, at about two in the morning. I was 20 to 30 kilometres north of Roma, travelling up to Townsville to see my son.

"There's a lot of little grey kangaroos along that stretch, so I was going very slowly, probably about 40 Ks an hour, out of respect for the wildlife.

"I'd just started to go into a forested area. There were a lot of larger trees on the left-hand side where the ground dropped away.

"I had my high beams on, and [could] see quite a distance ... and there was something [crouched on the road ahead] and I thought, 'That's quite big for a 'roo.' It was quite dark and big – so I slowed right down and as I was getting closer it stood up, and was just, like, staring at me.

"I initially thought it was about four feet tall, but it could have been four-and-a-half to five feet.

“As I was getting closer [its hair] was getting shinier – a reddy, auburny colour. Very shiny … maybe 10 centimetres long all over. It was just standing there, staring at me, and I was thinking, ‘Oh my gosh what’s that?’

“It was sort of looking at me straight on, but whether it was because it had its head down, or because the hair was so long … I couldn’t see any facial features.… there was no eye reflection at all.

“I seemed not to be able to see the feet – I couldn’t define a foot because, I suppose, it was facing towards me. It’s hard to define the shape of a foot with hair on it. I was looking at the overall colour and shape, because I’m a bit of an artist, and shape and colour come to mind first. So, no – I didn’t see the finer details and I didn’t look for them.

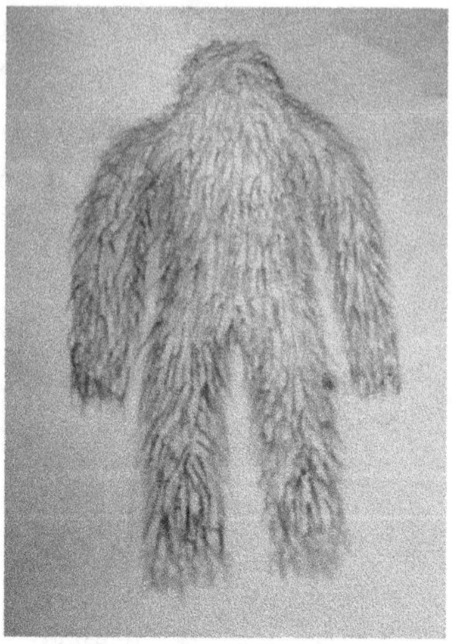

Mandy’s sketch

“The arms were long. It’s almost like it was leaning forward and looking down, so the hands were a little bit more forward [than they would have been if it was standing fully erect]. The elbows seemed slightly bent and the hands facing backwards – a bit like a tai chi stance, if you know what I mean – that sort of relaxed stance – the rounded shoulders. I couldn’t see any bare skin at all.

“I probably saw it for 40 or 50 seconds, because I slowed *right* down to about 15 Ks an hour. It was very close to me as I drove past. I had a good look, and I thought, ‘That’s just *amazing – just absolutely incredible.*’”

A very gentle, sensitive person, Mandy has a strong affinity for animals, particularly young ones. “I had the impression it was a young creature – you know?” [Just as one can sometimes tell if a dog or some other animal is young or old].

“It just stood there, didn’t move from side to side, and there was a little dead ’roo there, and I wondered if it was eating it.

“So, as I got closer, I thought, ‘I have to stop and take a photo of this’ but then thought, ‘Damn!’, because as I’d driven out of Roma, I’d turned my phone off. Then I thought, ‘If this is just ‘junior’ [there might be larger ones around] and I didn’t feel safe … so I kept driving.”

Perhaps because she’d had her windows wound up, Mandy hadn’t noticed any odour.

"I have a condition where my senses shut down, so I didn't feel the fear until a couple of days later [when] my body was in shock – it was, like, numb. So I started looking on the Internet, and I thought, 'So that's what I saw!' And then I got onto the [AYR] site and thought, 'I need to talk to someone about this.'

"It just confirms that it's real. I was blessed to be able to see it. It was just *extraordinary* – so *beautiful*, you know – such an amazing creature, and I thought ... and I had [also] seen the black panther, you know, in the Otways, with a friend and my sons."

Another black panther

Tony: "That's very interesting. My collaborator, Paul Cropper and I are very interested in the Australian big cat mystery – we included a couple of chapters about them in our first book. Could you describe that encounter for me?"

Mandy: "It was in 1999; I was driving from Torquay, inland to Anglesea, with my sons, my friend Chris, and her children. We were on Grossmanns Road, which is a bit windy and bushy, and we saw this big, black, sort of cat bound from the left-hand side to the right-hand side of the road, and I said, 'What's that?'

"It was *huge*, and Chris says, 'I think that's a panther!' The distance between its back legs and front legs was about half the width of the road. And its long tail went up and then came down [and] curled into a U-shape at the end.

"It was at least a metre a half, two metres long. I used to have a German Shepherd; it was bigger than that, lower to the ground, but longer [in the body].

"This was just before dusk, about 5:30 pm. We were about 100 metres from it and saw [its entire body] for about four seconds and then watched its tail as it disappeared.

"Most people in the area are aware of the creatures – they call it the 'Otway Panther.'"

Case 266. 29 Aug 2018. 10 km west of Dorrigo NSW

Lenny Keech contacted AYR on 30 August and was interviewed by Paul Cropper later the same day. Lenny's friend, Zoe Watts, also took part in the interview.

The terrain to the west of Dorrigo is very hilly, with numerous rivers, waterfalls, and large tracts of dense forest. The incident happened between 9.30 and 9.45 pm, as Lenny was driving back to Dorrigo after dropping Zoe at her property on Johnsen's Road.

Lenny: "Maybe two kilometres from [Zoe's] house, I came around a corner doing 90 kmh, onto a straight bit of road – it's quite hilly but you can see a good 200 metres ahead and, as I came to the start of the straight, I saw these yellow eyes.

"There was a full moon, so the silhouette of whatever-it-was, was quite visible, standing on its hind legs on the right-hand side – not on the road – up on a paddock. It would have been nine, ten foot. Yeah – it was *tall.*

"I travelled maybe 20 metres, and [then] it ran from that side of the road into a paddock on the other side, as quick as … he was *moving.*

"[As it sprinted] across … because the moon was so bright, I could see its silhouette. It covered about 100 metres in two seconds, maybe. It was *really* quick.

"You can see into that paddock, and as I drove past it was just looking at me, making sure I was [not going to stop]. It could have [run away] but it stayed there, looking at me.

"I could see pretty much every part of it; it was standing side-on, amongst three tree trunks. [By then] it was hunched over, [but even so] it was still six, seven foot tall. Its torso and arms were really long – real long arms hanging low towards its knees. It had a big chest.

"It was real hairy, black as the ace of spades – I could see hair on the back – long and hanging, like small dreadlocks – rough and matted [but] no hair on the chest that I could see. I had all the windows up, so there were no weird smells.

"It was massive … way too tall to have been an ape. You know the Prisoner of Azkaban in 'Harry Potter' – where he starts turning into a werewolf? [Laughs] A bit like that, but way more solid – bigger across the chest.

"The head was out of proportion – quite small. If anything, it didn't really have a neck. You know – just head-to-shoulders."

Paul: "Did you see much of its face?"

Lenny: "No – I could just see its yellow eyes – no face – the eyes sort of drew me into it. It made eye contact [and] these two yellow eyes, they were like nothing I'd seen – like two little suns. Big yellow eyes – the yellowest eyes you've ever seen – it was crazy – almost a devilish-type thing.

"I felt, I dunno, almost like I wanted to *yell*, if that makes any sense at all? But yeah – it was a weird feeling. I felt really uneasy afterwards, my skin felt cold, but hot – I felt real panicky, I just wanted to get home."

Lenny told Zoe about the incident after work the next day. "We went back there today and could see a definite track leading right across the road into the other paddock. Pretty big footprints in one spot – they would have been double my feet, and I'm size 10."

Paul: "Did you photograph them?"

Zoe: "We did, and because the grass is short at the moment, you can see them pretty well. The track looks well worn – like it's used a bit."

Lenny: "It's not somewhere you'd lead cattle through, or even a place where people would walk through – it's very steep country."

Yet another black panther

Paul: "Is there anything else you'd like to mention?"

Lenny: "Well, not really … I might have seen a panther … It was only three, four days before I saw [the yowie] but I don't know if there's any connection whatsoever.

"Again, I was driving Zoe home …"

Zoe: "I saw it as well! It was quite late, about 11 o'clock, and about 200 metres from my house – very much down in the bush. I have 100 acres [largely bush] and there's a lot more bush surrounding us."

Lenny: "This is only about a kilometre from where I seen the yowie … so that's why I thought there might have been a connection.

"It came up from some pretty dense bush on the left side of the road and ran straight across our path, quick as anything – faster than any wild dog. It was just jet black."

Zoe: "I just saw its bum and tail out of the corner of my eye, and Lenny said, 'Did you see that? And I said, 'That black thing?' And he said, 'Yeah!' He saw the whole thing, and said it was like a big panther, and from what I saw, it was.

"I talked to my grandma about it, and she's also got friends who've also seen a panther around our road – Johnsens Road."

As readers will have noticed, several other yowie witnesses have mentioned encounters with mysterious "black panthers". As only three or four days elapsed between his sightings of the mystery moggie and the hairy giant, it's not unreasonable for Lenny to wonder if there might have been a connection.

It's interesting, also, that the site of his yowie encounter is only three or four kilometres south-east of Deer Vale, where our friend Gary Opit investigated the remarkable "gorilla-men" episode in 2011 – 2012. (Case 246)

Case 267. 10 Sept 2018. Bunyip North, VIC

Jason Fry's sighting occurred on his large property on Wimpole Road, Bunyip North. He contacted AYR immediately afterwards and was interviewed by Paul Cropper the following day.

The gentle giant of Cannibal Creek

At about 7.30 am on 10 September, as he walked down a grassy laneway towards his horse paddocks, Jason's eyes were drawn to two horses standing very close to a fence that separates his property from Cannibal Creek.

> **Jason:** "There's a lot of trees and scrub along the creek; quite often you'll see a wallaby in amongst the horses.
>
> "There was something on the other side [of the fence] behind the horses – I could see its head and [right] shoulder. The horses are about 15 hands – about 1.5 metres tall – and this thing was head and shoulders above them. I was about 50 metres away and I could see it clearly from about eight inches below the shoulder up to the head and neck. I've never seen anything that tall and humanlike.
>
> "I only saw it, side-on, for maybe three seconds as it covered, probably, ten feet – three metres – from left to right, behind the horses … and out of my field of vision because of the tree guard [wind break].
>
> "It was definitely hairy, the same colour as a wallaby – dark brown, almost black, but it was not a kangaroo – it didn't hop or bound – just moved quite gracefully, an even pace, it wasn't running.
>
> "It was definitely a humanlike figure [but there was no suggestion of clothing]. Very, very tall. [It wasn't] heavy-looking – it was quite slender.
>
> "The hair was not too short but not shaggy, over the whole body [but] I reckon it didn't have hair on its ear [because] there was a shiny area, quite shiny and black, on the side of the head that I assume was its ear. I couldn't see whether it had a flat face, or a pointy nose.
>
> "I was pushing a wheelbarrow, so I just continued down the laneway, until I got to a gate on the right, where there's a gap in the tree guard. I popped my head out there, hoping to see it moving further along the fence line, but no – there was nothing there: gone."

When he looked over the fence, Jason realised the creature had been wading in the creek as it passed behind the horses.

> "It's a slow flowing creek, with a lot of weed in it, and I could see where something had passed through the water ... through the reeds and weeds, from one side other. [So] what I saw would have been about two feet taller than it first appeared – in the vicinity of eight foot.
>
> "The creek banks are quite dry. I looked for footprints but couldn't find anything."

He added one rather strange detail: "The horses weren't startled at all – one was actually grazing [as the creature passed] and the other one was looking right at it – and they were side-by-side [with it]. They weren't perturbed at all."

While the neck of the creature, as depicted in Jason's sketch, appears unusually thin, we must bear in mind that his sighting lasted for only three seconds at a range of 50 metres.

That detail, in fact, is one reason why we consider Jason to be an honest informant. Surely anyone attempting to perpetrate a hoax would have fashioned his or her sketch to conform to the extremely thick-necked yowies depicted in various articles, books, and websites.

His mention of the horses' apparent unconcern as the yowie strolled past will strike some readers as very odd. But while our files are full of cases in which horses have panicked at the merest hint of yowie activity, there are a few others in which they have remained strangely calm in the presence of the hairy giants. (e.g., Case 261)

Jason's sketch

Black panther – again

Something else is worthy of note. When Paul asked if he'd had any other unusual experiences, Jason said he'd once encountered a black panther.

It happened when he was about 16 years old, near his parents' property at Garfield North – "about five kilometres from here, and on the same creek."

He was walking home one night through a large bushland reserve when he realised, "I was definitely being stalked. There was bright moonlight, and I saw it – about waist-high and really long. It would have been over two metres long – and black as pitch. That cat's been there for [a long time]. We grew up knowing there's panthers in that area."

Over the last 140 years or so, that region of Victoria has certainly produced many reports of what appear to be super elusive, out-of-place big cats – mainly "black panthers." The "alien big cat" phenomenon is, in fact, one of our country's most intractable zoological mysteries – second only, in the weirdness stakes, to the yowie enigma.

We find it interesting that so many individuals seem to have encountered both types of cryptids – the mystery moggies *and* the hairy giants – not only in Victoria, but also, as we have seen, in several other parts of Australia (and in North America, for that matter.)

Bunyips

In this particular case, the odd conjunction of cryptozoological mysteries is even more pronounced – just a couple of hundred metres downstream from the site of Jason's yowie experience, Cannibal Creek joins the equally evocatively named Bunyip River.

Throughout the early colonial era, the existence of bunyips – large, super-elusive water monsters dreaded by Aborigines and reported on dozens of occasions by British settlers – was hotly debated by scientists and explorers. Although the creatures were written off as myth by about 1850, people have, on very rare occasions, reported seeing them well into the modern era.

Victoria, it might be noted, produced more Aboriginal bunyip lore and settler reports than any other region of Australia – hence place names such as Bunyip North and Bunyip River. (See Chapter Six, *Out of the Shadows*).

Case 268. 26 Oct 2018. 20 km south-west of Agnes Water, QLD

Stef van Wijchen, a 25-year-old Dutch backpacker, contacted AYR via email just six hours after his brief, but very interesting, sighting. As he was on the road, in the middle of his Australian adventure, we were unable to contact him for a chat, but his email – quoted in its entirety below – tells the story very well.

> "Right now, I'm in Agnes Water, Queensland. I just arrived here this morning by Greyhound bus.
>
> "While riding the bus, I woke up when the sun came up. Time of day: 6.00 am. Weather: sunny, clear sky. I really like to just look through the window because this land looks so much different than my own.
>
> "While I was looking out the window, I saw a big, hairy, humanlike appearance in a bit of a more open bush area, at a distance of about 100 metres, and it was walking. It looked kind of like the Wookie from Star Wars.
>
> "Now, I'm a pretty down-to-earth guy and I usually don't really believe in

folklore or anything like that. But this sighting freaked me out a little bit and got me interested. Have there been any known sightings of big ape-like humans or something like that in the past?

"PS: awesome country and excuse my English."

Stef certainly didn't need to apologise for his English, which, like that of so many of his countrymen, is very good.

As for his query about similar sightings – there have been at least two in the vicinity of his own encounter. On 24 April 2012 "C. R." and a mate saw a huge, "primate-like" creature just six kilometres to the south, and on 4 May 2014, Angel Owen sighted an eight-foot-tall specimen in Deepwater National Park, about 12 kilometres to the east. (Cases 247 and 248)

Case 269. 13 Nov 2018. Beechmont Road, near Canungra Land Warfare Centre, QLD

When 53-year-old Glenn Kilmartin contacted AYR on 17 Nov 2018 Dean Harrison immediately interviewed him and visited the location. Paul Cropper conducted a long interview with Glenn on 24 November.

Glenn: "I'm a truck driver. I work out of a quarry at Beaudesert delivering mainly road base.

"This thing scared the absolute crap out of me. It happened at about 10.10 am at Witheren, [travelling north] on Beechmont Road."

That stretch of road forms the western boundary of Canungra Land Warfare Centre, a vast, rugged area that has, for many years, been Australia's main jungle-training base.

"There's a big, ten-to-twelve-foot fence, so you can't get in … it's wild on that side – miles and miles of deep bush. On the left-hand [western] side it's sort of half-bush with ten-acre properties scattered around.

"I was coming down the hill empty, and started going around a sharp right-hand corner, and thought I saw a boulder rolling onto the road off the embankment, so I hit the brakes to avoid hitting this 'rock' – and it stood up!

"I managed to skid to a halt [the creature had been in the right-hand side gutter] but had already started to step out as I was still skidding, and it spun around to face me, so [when the truck stopped] it was standing right in front of my bonnet. It seemed shocked that I was there."

The enormous creature was standing so close to the truck that all Glenn could see at first was its upper body, from naval to shoulders. "He was a metre, metre-and-a-half … four-foot across the shoulders. I reckon he would have been [the weight of] three good-sized men, so … 400 kilos – that seems about right.

Glenn's sketch – "It was standing right in front of my bonnet."

"The truck is an International S-Line, a standard tip-truck – bonnet out the front, no bull bar – so it was lucky I didn't hit him, otherwise … I think he could have picked up the front of the truck – probably about three ton – I reckon he could have done that.

"It had hair probably two inches long all over its body; really dark brown with a reddish tinge, like an auburn tinge – not like rust – and matted. It had a naval – I could see through its hair. The top of my bonnet is six feet off the ground, so its belly button was six feet high.

"I had my sun visor down, so I had to scrunch down and duck my head to see its face. The head was as high as the top of the truck; that's nine, nine-foot-six, and he was a bit higher than that – ten foot, if it was an inch."

Glenn Kilmartin. (Dean Harrison)

It had "… a rounded head – it wasn't that conical head some people talk about. It looked like a chimpanzee's head [and it] looked too small for its body. No neck – like its head was on top of its shoulders [but] it seemed to be sort of pushed forward, like it was growing off its chest rather than off its shoulders."

The face was largely free of hair "from its lips up to its eyebrows – no hair on the top of its cheeks. It didn't have any hair on its ears, either.

"A flat nose – like a boxer who'd had his nose broken a couple of times. I didn't see any teeth – it had its mouth closed. It was dark around the eyes and its eyes were black, and the centre of the pupil was hazel coloured."

Paul: "Did you get the impression you were looking at an animal or a human?"

Glenn: "Animal. The face – I saw three different expressions: shock, embarrassment, then anger. It went wide-eyed with shock when it first saw me, furrowed its brow, and it was, like, 'Oh – I shouldn't be here!' or 'He shouldn't be here!'

"And I don't know whether I'm imagining it

Sketch by Buck Buckingham under the direction of Glenn Kilmartin

– but it seemed to be a little bit embarrassed; then its brows knitted deeper and became an angry expression. Those expressions flashed over its face – they were its only human [characteristics].

"Then it grunted – just a loud grunt, not a howl – a short, sharp grunt – and slapped ... well, punched, the centre of my bonnet, with his palm instead of his fist – like a push: 'Get outa my way!'

"It didn't move the truck much, but it shuddered the bonnet. Didn't do any damage [except for] a fine crack up in the bonnet's top left-hand corner; that's [because] the bonnet was pushed into the truck's body. But there's no mark where his hand actually impacted the front of the bonnet.

"But, oh yeah, I felt it – felt the push. It was almost as if I'd hit a small car or something – a definite shudder in the truck [which is] nine-and-a-half tonne empty.

"I saw his hand, but I didn't get the chance to count fingers, I just figured he had four fingers and a thumb, but I couldn't attest to that, because really, it was all over in five seconds – like, from start to finish.

"Then he just turned to his right and – two steps and he's gone – outa there – into the bush on the left-hand [western] side.

"Then I just rolled down the hill – didn't put it in gear – around the [bends] and pulled up at the bottom, and just sat there. I was shaking; rolled a smoke, and I couldn't believe it.

"I've never seen anything so big, so 'magnificent' is the word I'd use ... because he was just ... whatever he was, he was a *really* fine specimen – you could see his muscles underneath the hair; pecs and abs and all that. Everything about him was magnificent. He was an ugly sucker, but his body was just beautiful."

Because the creature was standing so close to the truck, Glenn didn't see any genitalia, but it seemed quite obvious it was male.

"And when he walked off, I could only see its rear end – and his bum was covered in hair, too."

Paul: "Did you talk about it to anyone?"

Glenn: "No, I didn't tell my wife until a week ago – I didn't want to get labelled a whacko. I was worried she wouldn't believe me, but in the end she did. She said, 'Well, I'm sure there's other stuff out there, too, that we haven't discovered.' And I told [the rest of] my family today. I decided, 'Well, I've got to get it off my chest.'"

A yowie "hot spot"

After his near-collision, Glenn checked out the AYR website and noticed that another (or the same) yowie had been seen crossing the same stretch of road on 10 March 2015.

While listening to a sound file of Paul's interview with that witness (Robert Cook, Case 252) and viewing Dean Harrison's film of the site, Glenn thought it likely his encounter occurred at "exactly the same spot."

In fact, two other people have reported very similar events in the vicinity of Canungra.

On 21 March 1986, a gigantic, long-legged, shaggy ape-man ran across Beaudesert Road, which passes through the Land Warfare Centre (LWC) about four kilometres north of the site of Glenn's experience. When interviewed by Paul, the witness involved in that event, soldier Lester Davidson, said that "The infantry fellows at 10 RIC had seen quite a few … in the bush in the LWC. Plenty of stories." (*The Yowie*, p. 84)

Three kilometres east of the LWC, 17-year-old Aaron Carmichael actually collided with a similar creature on 2 March 2001. His car, considerably damaged, was sent spinning off the road. Dean Harrison thoroughly documented that incident, photographing the damaged car, just a few hours after the event. (*The Yowie*, pp. 67-68)

Case 270. 2 Aug 2019. Deepwater, QLD

Aaron Nimmo contacted AYR on 8 Aug 2019 and was interviewed by Dean Harrison the next day.

Aaron lives on a 60-acre bush block close to the southern boundary of Deepwater National Park.

Since 2007, when he first moved onto the property, he has experienced a series of strange events, culminating in a sighting of a group of extremely tall, bipedal, non-human creatures.

On his very first night on the property, Aaron was startled by a series of very loud, thumps. Slightly unnerved, he locked himself inside his caravan.

The odd activity intensified after he moved a transportable cabin onto the block. As he was settling down to sleep on the first night, "Something banged on the [wall] right beside my head – scared the bejeezus out of me!"

Since then, he has regularly found the trunks of young, healthy trees cleanly snapped way above [human] head height; and branches of other trees broken and pulled down, as if to deliberately obstruct pathways. Logs have often been picked up and placed across tracks or stacked to create peculiar "tee-pee" structures.

Buck Buckingham

Buck is a well-established multi-media artist whose "day job" involves the creation of props for theatre, film and television.

He has been hooked on yowie research since the late 1970s, when, as a teenager, he sighted two of the creatures in south-east Queensland.

Since 2005, he has become a mainstay of Dean Harrison's Australian Yowie Research organisation, taking part in numerous expeditions and providing many eye-catching illustrations for the AYR website.

He is at his happiest when surrounded by trees, sketch pad and pencil in hand.

Finally, at about 6.30am on 2 August 2019, after hearing a "great commotion", accompanied by the sound of neighbour's dogs "going berserk, like an intruder response" he watched from his cabin door as a group of huge, dark figures ran at tremendous speed, north-to-south, through the nearby scrub. As the early morning light was behind them, he saw them only in silhouette.

They were upright, very tall – about eight feet – and twice as bulky as humans, and there several of them, probably about eight. Strangely, in view of the initial commotion, and although they seemed to be weaving their way between trees, Aaron detected little or no noise as they ran.

Although he scouted around for tracks after that event (as he'd done previously, after discovering tree damage) Aaron has never found tracks of any significance on his property. Interestingly, however, some years earlier, on a beach some distance away, he and a friend came across some remarkably large, bare-footed tracks: "They were quite massive, and we thought at the time they'd been made by a really huge man."

Just prior to his conversation with Dean Harrison, Aaron learned that yowies had been seen very recently on a property on nearby Matchbox Road. The landowner was in the process of setting up surveillance cameras to verify what she had seen.

The Deepwater area is something of a yowie hot spot. Aaron's property is about five kilometres south of where Angel Owen encountered an eight-foot hairy giant in 2014 (Case 248).

Case 271. 5 Oct 2019. Near Woy Woy, NSW

Duncan (surname confidential) contacted AYR on 15 Oct 2019 and was interviewed by Dean Harrison the following day.

> "I've been living in Woy Woy for 3 years. I'm an avid cyclist and have [recently] been doing a trail from the Woy Woy Tip to Mt. Wondabyne and across to Patonga.
>
> "For the past few weeks, there's been half a dead dog on the trail – just the back end, no torso or [front] legs. It was a bit strange. When I went up there on Saturday it was gone.
>
> "I headed off at about 9.30am, went up to the Woy Woy Tip and up to the right along Dillions Fire Trail. The bush became silent past where the dog [had been], no bird noises, a very slight wind and dead quiet. It was eerie … I felt that something was watching me.
>
> "At about 10.30 … I was on the Patonga Fire Trail and found a single trail to the left near Dillion's Farm and took that into the bush. The first thing I noticed were the trees [were] snapped, like something had been through there. It was very quiet and eerie. All of a sudden two trees to my left, at about ten feet up, started swaying. I just thought, 'Whoa, something's in there!' Then something moved and I felt uncomfortable, so I turned around and went back to the Patonga Fire Trail.
>
> "As I rode, I kept noticing this black shape off to the left and sometimes off

to the right. There were black shapes moving through the forest, following me parallel. It was really weird because normally you would hear the crunching of trees and stuff like that, but I couldn't hear a thing. They were moving like 'Predator' or something. It was really bizarre that these things were making no noise.

"There were times when I was walking, because there's ups and downs, and I don't like riding uphill.

"…these things [were] 50 metres off the trail and on either side at times … keeping pace with me. It got to me … I was that scared I had to stop and yell at this thing: 'I can f***ing see you watching me!'

"I kept riding, and again came to a spot where I felt uneasy and could see one to my right, behind a black tree stump. When I stopped, it stood up, turned, and looked at me. It was far bigger than me. I'm six foot six, and it was easily a couple of foot on me – perhaps nine or ten feet tall. The shoulders were about a metre wide. The head was very wide, the size of a big motorbike helmet. No neck … hunched neck. It was covered in black hair [that] looked half deadlocked … hard to tell because it was fifty metres away, but it was three to four inches long. It looked like an ape, like a gorilla, like a silverback. I saw its eyes look at me. It had a gorilla's face. The whole experience was so surreal … very bizarre.

"I absolutely sh** myself and was straight out of the fire trail and back on Patonga Road and raced home as fast as I could.

"I didn't know if it was someone in a gorilla suit playing games or what. I got onto my computer and googled the history of Dillion's Farm and then this can of worms [a history of yowie activity] opened up, which made me contact you guys. I've been obsessed since.

"I've ridden these trails dozens of times. I've ridden in bush locations from Melbourne to Cairns … but [have seen] nothing like this. It scared the bejesus out of me!

"I will be going back there. I would like to get some type of evidence."

Another black panther

Interestingly, when Dean asked if he'd ever seen anything else unusual, Duncan mentioned that (like *nine* other yowie witnesses in this collection of cases) he'd once sighted a black panther. "It was larger than a dog, on the Wooli Road, years ago."

Case 272. April 2020. Bellbird Grove, QLD

Immediately after the witness, an academic who requests anonymity, contacted AYR, he was interviewed by Dean Harrison and Sarah Bignell. A few days later, he escorted Dean and Buck Buckingham to the site of his encounter.

Bellbird Grove is just beyond the western outskirts of Brisbane, on the eastern edge of 1000 square kilometres of mountainous forest, bounded by Wivenhoe and Somerset Dams on the west, and extending north as far as Mt Archer.

The following was compiled from the witness's written submission to AYR and his interview with Sarah Bignell.

> **Witness:** "I've been walking these tracks since the 90s, and quite regularly – two to three times a week – for the last 10 years, taking photos of fauna, and had never seen anything strange in all that time.
>
> "I know I wasn't supposed to be walking through there that day, because the park was closed due to Covid – so I knew nobody was there. There were no cars in the carpark.
>
> "It was late afternoon, and I was on an inland mountain walk that was quite densely forested, following beside the creek. It was very quiet – the birds weren't even chirping; it was strange and hard to explain.
>
> "I could hear rocks clacking and echoing in the distance ... loud, you could hear it for a mile. There was a clack, pause, clack, clack, pause, clack – almost a rhythm to it. I thought there must have been someone further [up the creek]."

The track was more or less parallel with, and about 10 metres from, the creek.

> "So, I turned [off the track], and started walking up the bed of the creek towards the [sound] and ... came across what I first thought was a film set. Then I realised there were no cameras.
>
> "There was this thing [about 25 metres away] frolicking like a toddler in the ankle-deep water, acting in a really immature, infantile way, with strange movements – very primate. [It] was sort of twisting around, splashing, picking up rocks and throwing them, doing weird actions. Although it appeared to be young, it was still around six feet tall.
>
> "After about five seconds it turned around, and we both stopped and looked at each other. Then it raised its eyebrows, turned to the bank, pursed its lips, and made chirpy/kissy type noises.

"frolicking like a toddler in ankle-deep water"

"Then all hell breaks loose – this much larger, hairier creature stepped down into the creek from the left-hand bank and basically did an Incredible Hulk type pose as it was growling at me – an earth shattering, growling sound – the most reverberating yell, howl, growl, all in one.

A "much larger, hairier creature."

"I had two cameras on me, but I wasn't hanging around. I bolted – I'm gone!

"I'm a sprinter, and that was the fastest I've ever run. I backtracked along the creek and up onto the track, got to a clearing, bent over trying to breathe, got the camera out and just stood there trying to compose myself. I felt safer, made sure I had enough breath, and ran all the way back to my car.

"It's been playing on my mind ever since. I have a scientific background and if you had told me these things existed prior to the event, I would not have believed you.

"The big one was about two metres tall, a metre wide, and looked healthy. The face was gorilla-like, but different … rounder. The nose was wide and flat. The head [was] conical towards the back, not a human-shaped head. I think hair was covering its ears. [It was covered all over with] dark brown, shaggy hair, a slightly darker shade of brown than the smaller one, with a tinge of orange and a lot more hair … across the head, the forehead area and down the sides and back.

"The younger one was not an infant, perhaps a teenager in yowie years, and certainly not as wide as the larger one. It had shorter, stumpy legs, not athletic, and they were wider at the top, yet still fairly wide at the bottom, as if they were out of proportion. The same head shape. The arms were also out of proportion compared to a human's, as they were longer and top-heavy [and] bowed. They both had hair all over their bodies. No neck.

"The [big one's] facial expression was just angry. He had a scrunched up, wide-open mouth making that 'Get out there' growl. His teeth weren't fangs, but more pronounced, off white to yellow canines.

"It could easily have caught me. If it wanted to kill me, it would have. Why it didn't come after me, I don't know. I think its main job was to scare the crap out of me, and it did. It was fight or flight, and I was out of there, I just ran.

"It was life changing. What I stumbled across, lucky or unlucky, were beings unknown to science. It shook me to the core. It leaves you questioning everything. There's no way I would ever go there at night."

When Dean Harrison and Buck Buckingham visited the location with the witness, it became apparent that the larger creature had been sitting behind a tree from where he could monitor the walking track in both directions, while supervising the juvenile.

But while he'd been in a perfect position to see hikers on the track, he evidently wasn't expecting anyone to walk up the actual creek bed.

Under the witness's direction, Buck, a very talented artist, created a few excellent images of the hairy duo, which he superimposed on some of Dean's photographs of the site. They can be viewed, in colour, on the AYR website, www.yowiehunters.com.

Case 273. November 2021. Olney State Forest, west of Mandalong, NSW

The witness, Beau (surname confidential), contacted AYR on 25 December 2021 and was interviewed by Sarah Bignell on 12 January 2022.

Beau: "Me and a mate went dirt bike riding. We hadn't been to that forest before, but we went a long way in, really deep in – [beyond] any tyre marks … over a lot of obstacles, and I [went over a log] then pulled over and waited for my mate.

"Just as I stopped, I looked to the right and saw this big, like, blonde creature. [It was the same colour as the trees] And I'm looking, thinking, 'Is that a tree? Is that a cow, or a deer or something? 'Is that a human – a hiker or something?' But this is pretty far into the bush – we didn't see anyone else all that day.

Sarah: "What part of the creature did you see first?"

Beau: "It was so camouflaged with the environment that I thought it was a person in a ghillie suit until it started moving, and [then] I could see it wasn't any loose layer, and I wasn't sure until it moved, and I saw its whole head and arm.

"It was in line with all the trees and peeked its head around a tree, looking at me, and it had an arm wrapped around the tree – its whole [right] arm – I'm not sure if I saw its hand. It wasn't hiding itself very well! [Laughs]

"I was looking straight at it, staring at it for so long, and I had that feeling – like, 'fight or flight'. I thought it was gonna run at me or something, but it looked more scared of me than I was of it. I had no clue what it was.

"Then it slowly moved away from the tree, so I could see its entire body, and then walked backwards [with] its arms out a bit, crouching a little bit. All its movements were really slow … walking backwards, one step at a time, in 'stealth mode.' It almost looked like it was intimidated by my bike, because it was pretty loud.

"I was a good 15 to 20 metres away, but I could still see its shape and everything. It was a big, furred creature … really tall – huge, like eight or nine foot – really, really tall. So big and broad, but human shaped … fur all over – the hair was probably ten centimetres or so long and looked thick … a light brown/ blonde – a really light colour; beige, like…"

Sarah: "Sandy coloured?"

Beau: "Yeah – that sort of colour."

Sarah: "Could you see any eyes or mouth, or skin?

Beau: "No, I didn't see any of that – it was all hair. It was like it had no face – that's what weirded me out."

Sarah: "Did you notice any smell?"

Beau: "No – but I had my motorcycle helmet on. I wasn't sure whether to run or stop, so I turned my bike off, and was just *staring* at it. And I was turning my head and yelling at my mate to get over the log – and it just disappeared into the bush."

Sarah: "After you turned your bike off, did you hear anything – twigs snapping, etc.?"

Beau: "Well, no – my mate's bike was still going. He came over – I wish he'd seen it too – and he said, 'It's probably a deer.' And I said, 'It's no deer!' We immediately went after it – just walking – and it [had gone] really deep into the bush, down a slope. We couldn't see any footprints or anything. I don't know what it was doing in there. It was really damp, a weird environment, all rain-foresty, and I was a bit scared. Then I said, 'I'm not going any further!'" [Laughs]

Sarah: "Did you get the impression that it was male or female, or old or young?"

Beau: "I'm not sure – it was just this *big* creature. [It didn't look young because] it was so big and tall and … *bulk*.

"I researched it when I got home, and told my parents, and my dad said, 'Was it a yowie?' And I looked at [illustrations of yowies] and it was exactly that shape – so I called you guys."

****

Our catalogue of cases now stretches from the 1820s right up to the 2020s – from the early colonial era to the space age. Indigenous people, of course, know that the Hairy Man has walked this land for a heck of a lot longer than that.

No doubt he'll continue walking it, always just out of reach. We may never catch up with him – but it's fun trying.

APPENDIX

In the still of the night

As mentioned earlier, Queensland's Gold Coast hinterland has long been notorious for yowie activity.

Since the 1990s, Dean Harrison and his AYR colleagues have conducted many expeditions in the area, coming close to the hairy giants on several occasions, finding tracks, stick formations, and other tantalising evidence. Finally, one memorable night in early 2021, they captured convincing video evidence of the creature's existence.

During early April of that year, Dean, Gary Lynn, "Buck" Buckingham, Shannon Guthrie and Jacob Fellows had focussed their efforts on one particular location high in the mountains. It was a remote, disused fire trail that they dubbed "Stickland Track", for reasons that will soon become apparent.

The track is exceedingly difficult of access, but when they reached it for the first time on 1 April, after a long, very steep ascent, they were glad they'd made the effort – because awaiting them on the track were many stick formations of the kind they and other researchers have found on numerous occasions in areas of intense yowie activity.

Several large sticks, up to six feet in length, were standing upright in the middle of the track, jammed several inches into the rock-hard ground – a feat Dean and his friends – all fit, strong men – could not replicate. Nor could the sticks have simply speared into the ground after falling from trees. Sometimes there were five or six of them in a line. Next to one, on the track's hard surface, was a footprint-like impression measuring 16 and a half inches (42 cm).

A couple of additional 14 or 15-inch foot-shaped impressions were discovered in ground litter downhill from the track, plus what looked very much like knuckle marks – evidently left by one of the creatures as it scrambled up the very steep slope.

One of the vertical sticks

In plain view along the track were many other stick formations – twigs and branches bent, twisted, broken and arranged to form "tee-pees", crosses and various other odd shapes – all carefully placed where they couldn't be missed.

As they'd sweated their way, sometimes hand-over-hand, up the ridges that morning, Dean and his crew had the strong impression that were being watched. It seemed likely, therefore, that the stick formations were territorial markers of some kind. Hopefully, they didn't translate as, "Clear off, or we'll tear you limb from limb!" But at the very least, they seemed to be saying, "Yes – we know you're here!"

All of the formations were strange, but some were so downright mind-boggling that they bring to mind the phrase "high strangeness". These were very flimsy, two-to-three-foot-long sticks, not much more than twigs, which had been – impossibly, it seemed – rammed into the very hard ground without breaking. There they stood – defying rational explanation.

AYR has examined such inexplicable, ground-penetrating, flimsy sticks in other locations as far back as the late 1990s, and the authors have puzzled over identical phenomena in two NSW hot spots. Sceptics, if they've managed to progress this far through the book, will roll their eyes and laugh, but this phenomenon – impossible, crazy and pointless as it seems – is quite genuine. It has to be seen to be believed. The only way to replicate it would be to drill holes in the ground with a thin, metal rod before carefully inserting the sticks.

The site was unquestionably a yowie hot spot. So, on 4 April, after a second scouting expedition, the team returned with all their gear, prepared for an overnight vigil.

“Because of recent rain,” Dean recalls, “the leeches and ticks were out in force, so we had to sleep off the ground, in hammocks. The location had a creepy vibe – everything became quiet and still, and we felt something was watching us. After dinner, we began hearing noises around the camp.”

They’d been using a Flir thermal imaging camera for a couple of years, but in early 2021 they invested in a much more advanced device, the Guide IR50 – which proved to be a game-changer.

After a quick lesson from Gary on how to work the new camera, Buck walked about 60 metres from camp. Standing in the middle of the track, he began panning around, looking for unusual heat signatures in the inky-black forest.

At that point, all of his companions were back in camp. To keep radio communications as quiet as possible, every man wore ear pieces.

After about 20 minutes, Buck noticed a couple of large thermal images on the northern side of the track, and managed to film two large creatures which had emerged from foliage to take up a position behind a large tree about 20 metres away. One creature hugged the tree and peered around it, its head and one shoulder clearly visible as it stared straight at him. The other, though more obscured, appeared to be looking in the direction of the other camera team.

Then came the “money shot” – the first creature leaned out from behind the tree and bent down to, apparently, pick something up from the ground. In doing so, it revealed a clear silhouette of its head, upper body and one long arm.

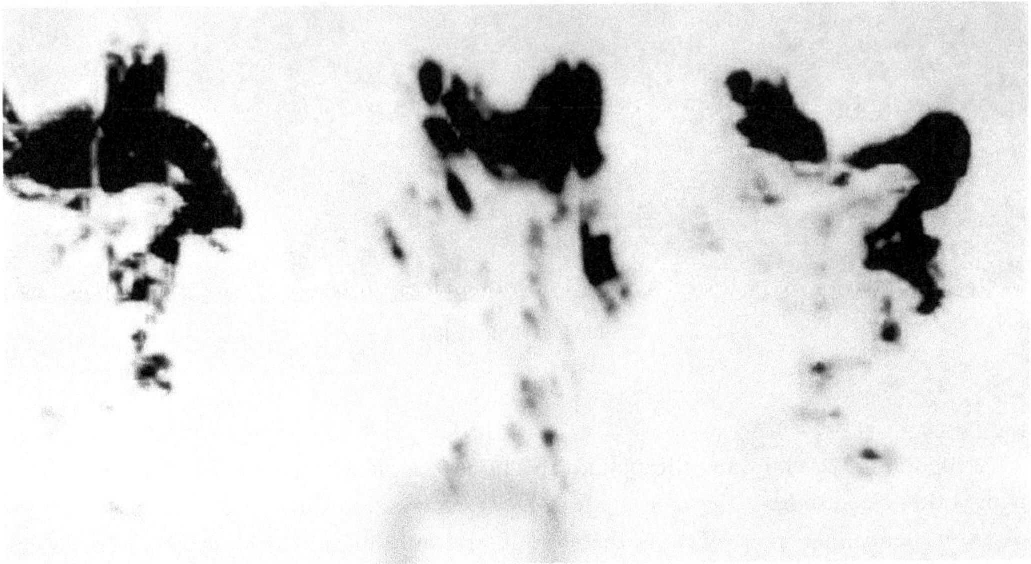

Three-shot sequence of thermal images

Buck: "I let the team know what I was seeing … and Gary said, 'I'm coming up to you, mate." At that, [both the creatures] looked in his direction and just took off – and that's it."

The creatures, though evidently very large, had made no noise whatsoever when they arrived at the site and none when they hurriedly left. If Buck hadn't been panning around, peering through the lens of the thermal imaging camera, he would never have seen them.

As Dean points out, the creatures could have been nothing other than yowies. The first creature's head is set directly on its broad shoulders and the silhouette of its head, in profile, is "classic yowie" – heavy brow, with a backwards-sloping forehead rising up to a high, egg-shaped skull.

There is no chance, either, that Buck was tricked by mischievous hoaxers. Not only is the site very remote and extremely difficult of access, but no human could move, in pitch darkness, through that tangled scrub without making a hell of racket.

AYR team examining images

No fear

As mentioned several times throughout this book, people who experience close encounters with yowies often suffer overwhelming fear. Dean experienced such terror himself during his first yowie encounter in 1997. It is therefore interesting that on this occasion, even though they knew at least two giant ape-men were lurking in the vicinity of their camp all night, none of the men felt at all threatened.

Dean: "We were up by 5 o'clock the next morning and walked towards where Buck got the footage. The previous afternoon and night there'd been no stick formations on that stretch of track, but now, exactly where he'd been standing, there were vertical sticks, and others laid out in patterns on the ground. One larger stick was right in the middle of the track, pushed in several inches.

"One bizarre-looking standing stick – roughly Z-shaped – wasn't the same sort of wood as the nearby trees and was too light to have dug into the ground naturally. Two other standing sticks formed an 'X' – a shape we knew all too well from other areas. There were no trees or branches above the X-marker from which the sticks could have fallen.

The Z-shaped stick

The X-shaped formation

"All of these signs were strategically placed just where Buck had been standing – placed so that we couldn't miss them. They'd been placed there between 1 and 6am.

"These new formations were the icing on the cake, but what did they mean? Were they a warning to their own kind that there were humans camping around the next corner – or a message to us, saying, 'We are here!' Or 'Get out! Or 'Welcome!'"

In daylight, it became apparent that the tree behind which the yowies had been lurking was 20 metres from the track. To accurately gauge the height and bulk of the creatures, Buck took several additional images, using the same camera, of team members standing next to the tree.

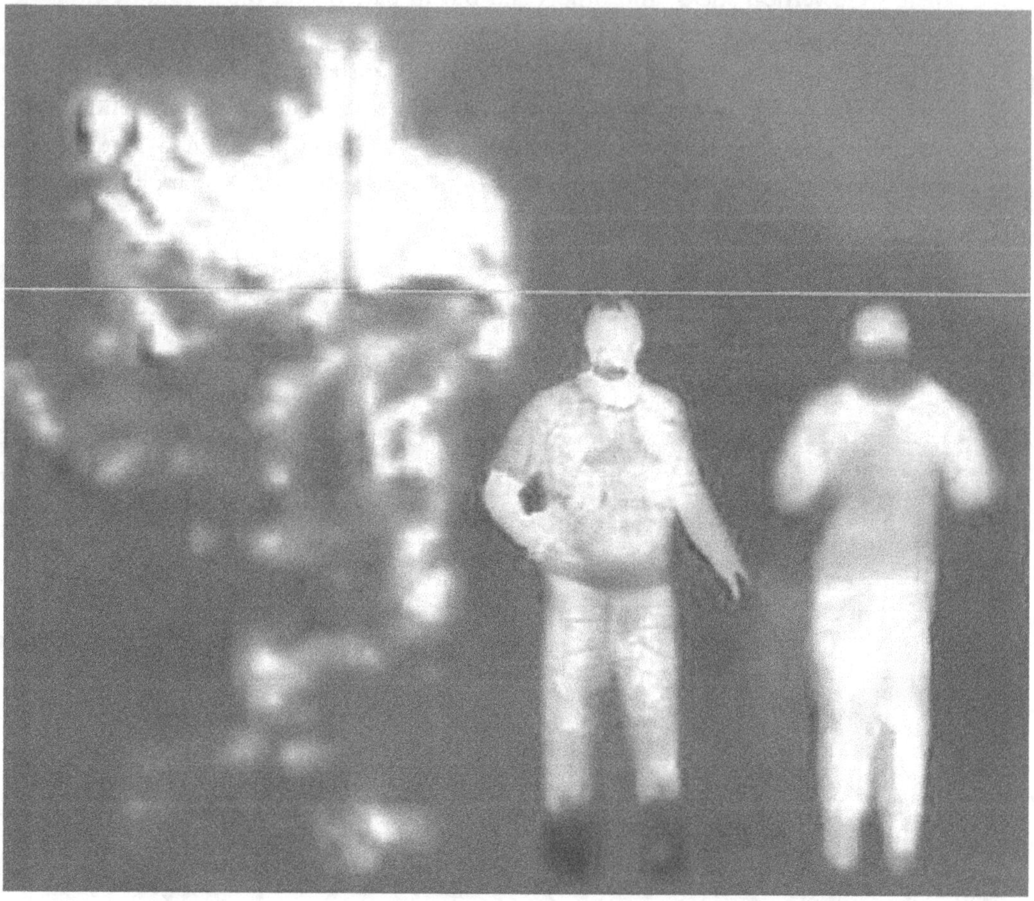

Thermal image of Gary and Shannon, both six feet tall, next to image of the yowie

Although Dean and his mates are thrilled to have obtained such clear images, they realise that the footage doesn't constitute irrefutable evidence of the creatures' existence. Sceptics will rush to say that the men were fooled by extremely stealthy nocturnal hoaxers – or that they faked the event themselves by filming a heavily-padded man standing on a box.

People who know and trust the AYR guys, as we do, don't doubt the authenticity of the shots, and we agree with Dean that the footage proves that the notoriously elusive Aussie ape-men *can*, in some circumstances at least, be photographed. The notion that the entire yowie phenomenon is a just an amalgam of hoaxes and hallucination is therefore no longer viable.

While the "Stickland Track" images may not convince everyone that the Hairy Man exists, we have no doubt they are genuine – and that they constitute important evidence.

We're sure that any open-minded person who goes to the AYR website, yowiehunters.com, and views Dean's video of the "Stickland Track" expeditions, which includes many images of inexplicable stick formations as well as the critical footage itself, will agree with us.

Dean acknowledges that although the film will be seen as an important breakthrough by many in the cryptozoological community, it is not going to set the wider world on fire.

"But we hope," he says, "that with these images we have, at least, brought some value to the table."

Log on a rock

On 7 July 2021, three months after the thermal images were taken, Dean, Buck, Garry and Jacob found another piece of very interesting evidence in a deep valley below Stickland Track.

> "There are no walking tracks in that valley", Dean notes, "and we've never seen humans there. We'd found stick formations and large, bare footprints that morning, but this was something else entirely – a large, dark scat with a sharp, overwhelmingly offensive smell."

"Large, dark ... overwhelmingly offensive."

It wasn't gigantic – just six inches long, and shaped like a human dropping, but its location – on top of a rock in the middle of a creek – seemed very strange indeed. "What human being," asks Dean, "would wade into a running stream to do their business on a rock barely above the waterline?"

Heavy rain had fallen most of the night, so the rock would have been underwater until the early hours of the morning. The foul calling card, therefore, must have been deposited after about 2 am.

And it was *really* foul: "Like nothing we'd ever experienced. Certainly not dog or human. As we bagged it, we were continually gagging and retching."

The relic, suitably frozen, was then forensically examined at Southern Cross University, Lismore. The scientist who conducted the tests noted that the scats of most ground dwelling Australian animals contain thousands of parasite eggs. Interestingly, however, this sample contained none at all. Humans and other primates, the scientist, added, are the only animals whose droppings do not have high egg counts.

> "This indicates," he said, "that the species that left the scat is surprisingly robust against parasites. Highly unusual and adds to the weight of the sample – looking good."

The scat was then sent for DNA testing at the Department of Primary Industries, Menangle NSW, where a Molecular Geneticist claimed he could not gather enough DNA to identify the animal concerned. That left the SCU scientist distinctly unimpressed: "They are claiming not enough DNA for an accurate identification," he told Dean, "[but] it was a substantial stool. My guess is they [simply couldn't] identify it."

BIBLIOGRAPHY

"A Squatter", *Reminiscences of a Sojourn in South Australia*, Kent and Richards. London, 1849.

Bicknell, Arthur, *Travel and Adventure in Northern Queensland*, Longmans, Green and Co., London and New York, 1895.

Clements, Nicholas, *The Black War*, University of Queensland Press, 2014.

Collin, Captain William, *Life and Adventures (of an Essexman)*, H.J. Diddams & Co., Brisbane, 1914.

Cutchin, Joshua and Renner, Timothy, *Where the Footprints End – High Strangeness and the Bigfoot Phenomenon*, Vols 1 and 2,

Derrincourt, William, *Old Convict Days*, T. Fisher Unwin, 1899, reprinted by Penguin, Ringwood, VIC, 1975.

Ellis, Netta, *Braidwood, Dear Braidwood*, N.N. & N.M. Ellis, Braidwood, NSW, 1989.

Favenc, Ernest, *The History of Australian Exploration from 1788 to 1888*, Turner & Henderson, Sydney, 1888.

Gale, John, *An Alpine Excursion*, serialised in the *Queanbeyan Observer*, 13 Feb to 17 Mar 1903.

Gordon, Stan, *Silent Invasion – The Pennsylvania UFO-Bigfoot Casebook*, Greensburg, Pennsylvania, 2010.

Green, John, *The Sasquatch File*, Cheam Publishing, Agassiz, BC, 1973.

Gregory, Denis and Manciagli, Alf, *There's Some Bloody Funny People on the Road to Broken Hill*, Snow Gum Books, Orange, NSW, 1993.

Hale, Horatio, *United States Exploring Expedition, During the Years 1838, 1839, 1840, 1841, 1842, Under the Command of Charles Wilkes, U.S.N. Ethnology and Philology*, Lea and Blanchard, Philadelphia, 1846.

Harris, Alexander, *An Emigrant Mechanic, Setters and Convicts, or Recollections of Sixteen Years Labour in the Australian Backwoods*, London, 1847, reprinted by Melbourne University Press, 1969.

Healy, Tony and Cropper, Paul, *Out of the Shadows, Mystery Animals of Australia*, Ironbark/Pan Macmillan, Sydney, 1994.

Healy, Tony and Cropper, Paul, *The Yowie, In Search of Australia's Bigfoot*, Anomalist Books, San Antonio, Texas, 2006.

Healy, Tony and Cropper, Paul, *Australian Poltergeist*, Strange Nation, Sydney, 2014.

Joyner, Graham, *The Hairy Man of South Eastern Australia*, Union Offset, Canberra, 1977.

Lapseritis, Kewaunee, *The Psychic Sasquatch,* Blue Water Publishing, North Carolina, 1998.

Lawson, Henry, *Triangles of Life and other Stories*, The Standard Publishing Company, Melbourne 1913.

McAdoo, Martin, *If Only I'd Listened To Grandpa: Recollections Of The Old Days In The Australian Bush*, Lansdowne Press, Sydney, 1980.

Massola, Aldo, *Bunjil's Cave: Myths, Legends and Superstitions of the Aborigines of South-East Australia*, Lansdowne Press, Melbourne, 1968.

Meredith, Mrs. Charles, *Notes and Sketches of New South Wales during a Residence in the Colony from 1839 to 1844*, London, 1844.

Murphy, Christopher, *Meet the Sasquatch*, Hancock House, Surry, BC, and Blaine, WA, 2004.

Nasby, C., *The Aborigines of Australia; stories about the Kamilaroi tribe*, as told to John Fraser, Maitland Mercury Office, 1882.

Pilkington, Doris (Nugi Garimara), *Rabbit Proof Fence*, University of Queensland Press, 1996.

Powell, Thom, *The Locals*, Hancock House, Surry, BC, and Blaine, WA, 2003.

Riddle, Tohby, *Yahoo Creek, An Australian Mystery*, Allen and Unwin, Crows Nest, NSW, 2019.

Robinson, Roland, *Black-Feller White-Feller*, Angus and Robinson, Sydney, 1958.

Smith, Malcolm, *Bunyips & Bigfoots, In Search of Australia's Mystery Animals*, Millennium Books, Alexandria, NSW, 1996.

Telfer, William, *The Early History of the Northern Districts of New South Wales* (also known as *The Wallabadah Manuscript*), c. 1900, University of New England Archives A147/V213.

INDEX

A

B

C

D

E

F

G

H

I

J

K

L

M

N

O

P

Q

R

S

T

U

V

W

Y

Z

www.ingramcontent.com/pod-product-compliance
Lightning Source LLC
LaVergne TN
LVHW062115200926
848484LV00007B/1675